# CHARTING NORTHERN WATERS

# *Charting Northern Waters*

## *Essays for the Centenary of the Canadian Hydrographic Service*

EDITED BY
WILLIAM GLOVER

Published for the Canadian Nautical Research Society
and made possible with financial support from CARIS
by
McGill-Queen's University Press
Montreal & Kingston · London · Ithaca

ISBN 0-7735-2710-9

Legal deposit second quarter 2004
Bibliothèque nationale du Québec

Printed in Canada on acid-free paper that is 100% ancient forest free (100% post-consumer recycled), processed chlorine free.

This book has been published with the help of a grant from CARIS.

McGill-Queen's University Press acknowledges the support of the Canada Council for the Arts for our publishing program. We also acknowledge the financial support of the Government of Canada through the Book Publishing Industry Development Program (BPIDP) for our publishing activities.

---

**National Library of Canada Cataloguing in Publication Data**

Charting northern waters : essays for the centenary of the Canadian Hydrographic Service / edited by William Glover.
Includes bibliographical references and index.
ISBN 0-7735-2710-9
1. Hydrographic surveying – Canada – History. 2. Canadian Hydrographic Service – History. I. Glover, William. II. Canadian Hydrographic Service. III. Canadian Nautical Research Society.
VK597.C32C44 2004　　526.9'9'0971　　C2003-905983-9

---

Typeset in Sabon 11/13
by Caractéra inc., Quebec City

# *Contents*

# *Maps and Illustrations*

# *Foreword*

A.D. O'CONNOR
*Dominion Hydrographer*

In the early 1960s, while in a merchant vessel on passage down the West Coast of Africa and studying our chart in the middle watch, I came upon a pecked coastline bearing the legend "Unsurveyed." I wondered how would one survey, and chart, such an area. That thought was to result in a career of some forty years in hydrography; hence my interest in this publication.

*Charting Northern Waters* is not a history of the Canadian Hydrographic Service (CHS), but it is inexorably linked to it. This fine compendium, skilfully pulled together by Dr William Glover, describes the early days of hydrography in Canada and includes the contributions of others in our hydrographic neighbourhood to the science. The CHS, one of the most modern, effective national hydrographic offices in the world, has evolved from the foundation laid by those whose exploits are recorded here. Canadians will be interested in these narratives, which also describe how the CHS predates what has become the Royal Canadian Navy by six years and the Canadian Coast Guard by sixty. Mentioned in passing are the Fisheries Protection Service and Canada's first steps into physical oceanography and other marine sciences.

I am struck by the similarities of the challenges faced by our earlier colleagues to those we face today: an expanding client base, increased demand for hydrographic information, a paucity of resources, yet a relentless determination to get the job done. Some things have not changed. Others have: advances in technology have increased our capability and, concurrently, the expectations of our clients. Meanwhile, let us not forget that many Canadian waters remain uncharted.

# *Foreword*

DR S.E. MASRY
*President and* CEO, CARIS

Understanding the past gives us a deep appreciation as to how far we have come and certainly helps us to prepare and build for the future. *Charting Northern Waters* provides an excellent insight into the trials and tribulations that faced the hydrographers of the twentieth century and earlier in their attempts to open the Canadian North for commerce and trade. The mobilization of Malaspinna's survey would seem unbelievable in today's fast-paced and modern hydrographic surveys, not to mention the gruelling task of performing the survey itself.

The chapter on the hydrographic survey work of Canadian government departments presents a kind of déjà vu when we read, "The ports of Halifax, Saint John, Quebec City, and Montreal and several Great Lakes ports were all slated for massive expansions." Is this history? This section goes on to describe the work required for the establishment of "more transport infrastructure to get goods from inland to the coast and from there through ocean ports." It is interesting to see that such industrial drivers exist today, with the added requirement of compliance to standards and interoperability.

CARIS is honoured and appreciative of being mentioned in the chapter on Canadian technological advances. We have been quite fortunate to have had the opportunity of working closely with the Canadian Hydrographic Service and other hydrographic offices in contributing to their technological initiatives over the past twenty-five years. As with all companies, we appreciate the opportunities and challenges that hydrography has presented us over the past several decades. The enduring character of many accomplished Canadians, through their innovation, perseverance, and commitment, has brought us to where we are today.

I certainly hope you will enjoy this in-depth look at how hydrography has progressed as we prepare to build for the future.

# *Acknowledgments*

Many people have helped along the way as this book progressed from idea to published form. The first individual who must be thanked is Tony O'Connor, the Dominion Hydrographer. The initial idea had been to hold a conference to celebrate the centenary of the Canadian Hydrographic Service. It was Tony who asked about a book and having launched the idea, gave it strong support along the way. In his absence, Dave Pugh of the CHS has been an equally avid supporter. But a book remains an idea unless it is written. An edited volume such as this has many authors and there would not be a final product unless each of the ten contributors had agreed to take part in the project and then met what must have seemed to be impossibly short deadlines. Chris Johnson has produced maps on an even tighter time line.

In my work on this volume I am pleased to thank Jim Pritchard for his constant support and advice, Art Collin and Ross Douglas, both former Dominion Hydrographers, for their assistance and suggestions, and Roger Martin, Elizabeth Hulse, and Joan McGilvray, all of McGill-Queen's University Press, for their work to produce the book to the same imposing deadline. I extend thanks to those unnamed who provided suggestions for authors in the early stages and to the anonymous readers who read part or all of the manuscript and offered valuable suggestions. The remaining deficiencies are my responsibility alone.

Finally, this book was made possible by the support of several organizations and companies. First, it is a pleasure to acknowledge the assistance of the Canadian Hydrographic Service in arranging corporate sponsorship. Dr Sam Masry, president and chief executive officer of CARIS, has generously stepped forward. Indeed, the centenary year of CHS also marks the silver anniversary of CARIS, a company that has always been associated with hydrographic work. So it is a special pleasure to congratulate and thank them for their participation in this project. And lastly, I gratefully acknowledge the support of the Canadian Nautical Research Society and of all my colleagues there.

# *Contributors*

CHRISTOPHER ANDREAE is the president of Historica Research Limited, a heritage planning firm he has owned for twenty years. He has a master's degree in industrial archaeology from the University of Birmingham in England and is currently preparing a PhD thesis on the development of the Port of Quebec in the nineteenth century. In 1997 he published *Lines of Country: An Atlas of Railway and Waterway History in Canada.*

GEORGE BOLOTENKO was born in Austria in December 1946 and immigrated to Canada in 1951. He received an MA in 1972, followed in 1979 by a PhD in Russian and European History from the University of Toronto. He then taught at University of Toronto and Ryerson for two years before joining the National Archives of Canada in 1981. Dr Bolotenko is currently an adjunct professor at the Centre for Russian-Canadian Relations at Carleton University. He has published in the fields of archival practices, Canadian history, and Russian-Canadian relations.

ANDREW S. COOK is archivist for the East India Company map, geographical, and maritime collections in the British Library. His PhD thesis was a bibliographical study of hydrographer Alexander Dalrymple. Dr Cook is currently investigating the development of coastal topographical knowledge for mariners and compiling historical catalogues of Admiralty sailing directions and charts.

ANDREW DAVID served for more than forty years in the Royal Navy, during which time he specialized in hydrographic surveying. Between 1961 and 1985 he worked in the United Kingdom Hydrographic Office, where he developed an interest in the history of exploration. Among his many publications is *The Voyage of HMS Herald to Australia and the Pacific, 1852–1861* (1995). Commander David was also the chief editor of the Hakluyt Society's acclaimed *The Charts and Views of Captain Cook's Voyages* (3 vols., 1988–97). He is currently one of the editors of the society's *The Malaspina Expedition, 1789–1794: The Journal of Alejandro Malaspina*, the first two volumes of which were published in 2002 and 2003.

RICHARD H. GIMBLETT served in the Canadian Navy and co-wrote the official account of the Gulf War. Now an independent historian, Dr Gimblett is a research fellow with Dalhousie University's Centre for Foreign Policy Studies, is engaged in writing the official history of the Royal Canadian Navy, and is vice-president of the Canadian Nautical Research Society.

WILLIAM GLOVER is a retired naval officer and an independent historian. He has written about the Royal Canadian Navy in the twentieth century and on aspects of the history of navigation. Dr Glover is currently editor of *The Northern Mariner/Le marin du nord*. He is a former president of the Canadian Nautical Research Society and a vice-president of the International Commission for Maritime History.

DAVID H. GRAY, who holds degrees in science and engineering, joined the Canadian Hydrographic Service in 1971, where he is currently the Geodesy, Radio Positioning, and Maritime Boundary Specialist. He is the author of papers on radio propagation, maritime boundaries, toponymy, and geodetic positioning, and he contributed the hydrography chapter to *Mapping a Northern Land*, describing the events of the post–Second World War era in the Canadian Hydrographic Service.

MICHAEL L. HADLEY is professor emeritus of Germanic studies and associate fellow in the Centre for Studies in Religion and Society at the University of Victoria. He is also a senior member of Robinson College, Cambridge, and a fellow of the Royal Society of Canada. A retired naval officer, Dr Hadley is recognized internationally for his prizewinning works on naval history.

JAMES PRITCHARD is professor emeritus of history at Queen's University, Kingston. He is the author of *Louis XV's Navy, 1748–1762: A Study of Organization and Administration* (1987), *Anatomy of a Naval Disaster: The 1746 French Expedition to North America* (1995), and *Elusive Empire: The French in the Americas, 1670–1730* (2003). He is also the president of the Canadian Nautical Research Society.

G.S. (STEVE) RITCHIE was educated at Dartmouth Royal Naval College going to sea as a midshipman in 1932. Five years later he joined the Surveying Branch of the Royal Navy, in which he served for thirty-five years. He commanded four surveying ships and served as Hydrographer of the Navy from 1966 to 1971. Rear Admiral Ritchie was elected president of the Directing Committee of the International Hydrographic Bureau in 1972 and 1977 for five-year terms of office. He has written several books on hydrography, including *The Admiralty Chart: British Naval Hydrography in the Nineteenth Century* (1967, 1995).

VLADIMIR SERGEEVICH SOBOLEV, who was born in Tashkent in 1948, holds master's and doctoral degrees and is an academician of the Russian Academy of Natural Sciences. After completing his archival studies at the Moscow State Historical-Archival Institute in 1971, he served as director of the State Archives of Kostroma Province, director-general of the united museums of Kostroma Province, and director of the St Petersburg Archives of the Russian Academy of Sciences. Since 2001 he has been director of the Russian State Archives of the Naval Fleet in St Petersburg. Dr Sobolev is the author of some seventy academic works.

GARY E. WEIR is head of the Contemporary History Branch at the US Naval Historical Center in Washington, DC, and an adjunct professor at the University of Maryland University College. Dr Weir is the author of many books and articles, including "Fashioning a Professional Dialogue in Oceanography: The U.S.

Navy and the Ocean Science Community, 1924–1960," which appeared in *Earth Sciences History*, and *An Ocean in Common: Naval Officers, Scientists, and the Ocean Environment, 1919–1961*, which was co-recipient of the 2002 Richard W. Leopold Prize.

CHARTING NORTHERN WATERS

# *Introduction*

WILLIAM GLOVER

The consequences of marine accident and disaster may take many forms. That a volume of essays should be prepared to mark the one hundredth anniversary of an agency spawned by accident is perhaps one of the more unusual outcomes. Yet the such is the present case. The Hydrographic Survey of Canada was created by Order-in-Council in March 1904, by amalgamating agencies of three different departments of government sixteen months after the *Sicilian* ran aground in the St Lawrence River. But this was not the first time a marine accident had exercised such an influence on Canadian hydrography. The loss of the *Asia* in Georgian Bay thirty-five miles from Parry Sound on 14 September 1882 had led to the establishment the following year of the Georgian Bay Survey. This was one of the three surveys amalgamated by the 1904 Order-in-Council. Since 1927 the combined Hydrographic Survey of Canada has been known as the Canadian Hydrographic Service (CHS). In 1983 CHS marked the anniversary of its direct antecedent with the publication of *The Chartmakers: A History of Nautical Surveying in Canada* by Stanley Fillmore and R.W. Sandilands. The centenary of the unified organization is being observed in 2004 with a conference and this volume. Should the name "Canadian Hydrographic Service" survive government reorganizations, amalgamations, rationalizing, and downsizing until 2027, that too will surely be cause for a centenary celebration.

This volume does not pretend to be a history of CHS. Rather, it is a Festschrift. While the emphasis is placed on events of the last century, the antecedents of the organization are remembered with articles about the hydrographic work of each of the European imperial powers that were concerned at one time with the nature of the Canadian coast. Nor can the twentieth-century work of other countries be ignored. The sea is by its very nature international, and as Canadians look north, they are frequently aware of their northern neighbours, particularly the United States and Russia. It is therefore both interesting and useful for comparative purposes to include chapters that look at aspects of northern hydrographic work carried out by other countries.

Individually, the essays have much to offer. James Pritchard provides a close look at early hydrography in New France. His era ends with the Seven Years War. It was during that war that a master of a British warship, James Cook, first came to notice with his accurate charting of the St Lawrence River and of then the west coast of Newfoundland. The period of Cook's great work coincided with and benefited from two important navigational revolutions – the ability to determine longitude either by chronometer or by lunar distances and the invention of the station pointer, which permitted the plotting of angular relations of points of land without reference to a magnetic compass, whose errors were still imperfectly known.[1] Cook's surveying standards must be considered a third great navigational revolution. The period examined by Pritchard serves as the foil against which the achievements of Cook, Malaspina, and La Pérouse may shine. For example, throughout the seventeenth century the correct length of a minute of arc of latitude, a nautical mile, was still a subject of great debate. Although in 1636 Richard Norwood had published his determination that it was 6,120 feet, which for convenience might be rounded down to 6,000 feet, over a century later its common use was by no means certain.[2]

But foil or not, the Seven Years War ended over two centuries of French navigation and hydrography in what is now Canada. The parallel to the observation cited by Pritchard that the charts of the Nova Scotia coast around Cape Sable were useless is found in Samuel Pepys's minute of his 1683 voyage to Tangier: that when land was found unexpectedly, the solution was to discard the chart previously thought to have been the best.[3]

Within the context of that standard, Pritchard recounts how the needs of the navigator were met in New France. He also introduces the problem of training hydrographers, a problem that is no less relevant today.

James Cook was a great navigator with strong connections to two of Canada's coasts, whose work might normally be included in this volume's section on colonial hydrography. Yet it is because his work is so well known and extensively discussed that it has been omitted. By contrast, the work of Alejandro Malaspina, a near contemporary of Cook's, has been largely overlooked, for reasons that Andrew David provides. Indeed, many Canadian readers east of the Rocky Mountains may be unfamiliar with the extent of the Spanish presence on the Pacific coast in the eighteenth century, remembered today by the Spanish names. Malaspina's careful selection of instruments, the attention paid to the length of a nautical mile, and the corresponding knotting of the log-line all reflect a change in attitude towards hydrography from the period reviewed by Pritchard. Another important difference that must not be overlooked is that Malaspina and his colleagues were naval officers, not academicians. David's discussion of Malaspina's determinations of longitude attest to the Spaniard's great care in a work that has stood up well, despite the significant technical advances in surveyors' instruments. This chapter will be a valuable contribution to the English-language literature on an important hydrographer.

Andrew Cook provides the chapter that is intended to reflect the British colonial period of hydrography in Canada. His work neatly answers the comment made by Pritchard that perhaps a chart was of little value to the seventeenth-century mariner. The extension of charting on the British Columbia coast offers a window on the organization of the office of Britain's Hydrographer of the Navy. This administrative and organizational historical theme is continued in subsequent chapters.

The two chapters dealing with the amalgamation of previously existing agencies as the Hydrographic Survey of Canada both address this theme. Richard Gimblett examines the amalgamation at the political level, hitherto ignored by both political and hydrographic historians. The nation-building objectives ascribed to the new service in the larger political context were as relevant at the end of the twentieth century as they were at its beginning.

Christopher Andreae examines another area that has been overlooked, namely, the hydrographic agencies of the other government departments that were part of the 1904 amalgamation. In addition to looking at the record of administrative and organizational problems that the new dominion government had to address, he examines the broader nature of hydrographic work in Canada. Problems of river water flow and water use may not immediately appear to be hydrographic. He points out, however, that when rivers were used for navigation, competing uses such as timber slides or power had to be reconciled. To do so, information and data was needed, as it also was for the design of harbour improvements and for the safe navigation of ships. These requirements indeed created a "grey" area for administrative jurisdiction.

Rear Admiral G.S. Ritchie's chapter about HMS *Challenger* on the Labrador coast forcefully reminds us of the real nature of hydrographic surveying. The stranding of a ship, the impact of the weather, and the remoteness of the region were all important influences on surveying work. The replacement of sail with steam power for surveying vessels had not materially altered the problem of supporting surveyors that was discussed by Pritchard and David. Ritchie's chapter also provides us with the human element of the work.

This chapter, together with the following one, raises the common difficulty of the spelling of geographic names. English orthography changes, and readers will notice that the names provided on the 1934 map of *Challenger's* voyage are spelled differently in the text. The latter uses the spelling of local places found in the *Sailing Directions: Labrador and Hudson Bay*, sixth edition (1988). Likewise, the English transcription of the Russian place names found in Vladimir Sobolev's chapter about the Northern Sea Route conforms to the spellings used in *The Times Atlas of the World*.

Russia's Northern Sea Route from Murmansk to Cape Dezhneva, the eastern tip of Russia extending into the Bering Strait, is well over twenty thousand nautical miles in length. It has been an important domestic shipping route and an alternative to railways and roads where and when they have existed. The Canadian sea route that may be offered as a very approximate analogy is the shipping route through Hudson Strait to Churchill, Manitoba. The distance from that port to the eastern entrance of the strait is 943 nautical miles.[4] When the account of the Russian achievements in charting, marking, and lighting the Northern Sea Route is read, it should be remembered that the navigation aids found on the passage to Churchill are only at the port. The *Sailing Directions* warn that "only a small part of Hudson Bay and Strait has been surveyed to modern standards ... In Hudson Bay, charting is generally at a small scale [covering a large area] except for a few areas. The east and west sides are primarily a collation of track soundings."[5] Vladimir Sobolev's discussion of charting along the Northern Sea Route in the first half of the twentieth century, translated by George Bolotenko, is a welcome addition to the limited English-language literature on the subject. His focus is on the European North, and he relies heavily on the archival sources. To the themes already identified of weather difficulties and the challenges of geographic remoteness, he adds the problems of political and administrative upheaval far beyond anything suggested by Gimblett or Andreae. The achievements recorded in the annual reports of the parties working on Russia's northern coast provide a benchmark against which the impact of the technical improvements discussed by David Gray may be measured.

With Michael Hadley's chapter we return to Atlantic waters and even the Labrador coast. However, his subject is the use of hydrography in support of clandestine operations of war against Canada. He also introduces the closely related field of oceanography. Changing temperature, salinity, and density created control problems for U-boats in their ascent or descent and detection problems for surface warships whose sonar transmissions were reflected off the impervious layer. As Hadley notes, the Canadian response to the difficult tactical situations of inshore anti-submarine warfare was to begin limited study of the oceanography of the area. He concludes that the program of broad-based research and technical innovation pursued after the Second World War was the correct Canadian response to the tactical challenges of the physical oceanography. In fact, the emerging emphasis on oceanography and the advent of new technologies developed during the war meant that hydrography was irrevocably launched on a process of fundamental change.

The Joint Committee on Oceanography was established in 1946.[6] It was an interdepartmental agency of the federal government, with invited membership, including CHS.[7] This was an important first step in broadening the field of hydrography through the application of other sciences and technology. More opportunities were to follow. On 29 April 1958 the United Nations Conference on the Law of the Sea concluded at Geneva with the signing of four conventions. One of these, the Convention on the Continental Shelf, defined the shelf as "the seabed and subsoil of the submarine area adjacent to the coast but outside areas of the territorial sea to a depth of 200 metres or, beyond that limit, where the depth of the superjacent waters admits of the exploitation of the natural resources of these areas and the seabed and subsoil of similar submarine areas adjacent to the coasts of the islands." More important, "the coastal state [was] given sovereign rights over the continental shelf for the purpose of exploring it and exploiting its natural resources."[8] Unfortunately, Canada was dependent on the goodwill of the Soviets and the Americans to tell them where the Canadian polar continental shelf was. In what must be a rare example of quick government action, even before the conventions had been signed, on 5 April 1958 the government established the Polar Continental Shelf Project.[9] Initially, hydrography was one of the prime areas of research, in conjunction with oceanography, geophysics, and submarine geology. The nature of the work and the challenges that had to be overcome could not be resolved using the methods of Malaspina or HMS *Challenger*. The necessary changes, however, were not achieved without difficulty. The experience of the Polar Continental Shelf Project helped to make clear to the CHS senior leadership that the hydrographer of the future had to be better prepared. That meant better educated. Beginning in 1969, the CHS launched a program that would send four hydrographers a year to university to acquire degrees in mathematics, physics, and geology. Other subjects were added as new needs became apparent. Many of the people responsible for the innovations and technological advances described by David Gray were involved with the Polar Continental Shelf Project and also took advantage of the university training program when it was offered. Another important benefit of these developments was the forging of links with universities.

David Gray has written an account of ingenuity and advance over the past fifty years. Readers who have first-hand experience of Seymour Narrows in British Columbia's inside passage can only marvel at the idea of measuring its current by observing the movement of milk bottles with flashlights inside them. And yet no yachtsman or mariner would want to be without the tidal atlases that were the ultimate product of such pioneering work. Gray's discussion of the difficulties of the Decca navigation system also serves to remind us that problems of geodesy, which Pritchard raised, remain to be resolved. This is a theme that provides continuity across these chapters.

Finally, Gary Weir returns us to the human scale of great achievement. Through William Metcalf's diary, which records high seas that forced the cancellation of work, novices whose enthusiasm wrecked havoc, and bureaucratic struggles for notice and priority at higher levels, we are again reminded of the environment within which all the work discussed here has been conducted. Weir's essay also offers a useful comparison with the Russian and Canadian northern endeavours.

The year 2004 marks the centenary of a political decision to respond bureaucratically to immediate problems. Yet as these essays demonstrate, charting northern waters is a human endeavour in the face of many adversities, which employs ever-refining science for a useful purpose. Such enterprise is not limited to any specific span of time.

# 1

# *Hydrography in New France*

JAMES PRITCHARD

In France the term *hydrographie* lends itself to confusion because it is a synonym for navigation and is used to denote a nation's riverine system. Much in hydrography is also included in oceanography, the science of the physical and biological properties of oceans. Even so, hydrography, or the science of describing, measuring, and charting bodies of water, appeared early in New France. Many firsts might be claimed from examining the history of early hydrography in New France, but the way that hydrographic science was pursued and organized was as important as the brilliance of its practitioners in explaining the mixed results of marine science in Canada during the French regime.

Early ambitions to exploit the indigenous people of North America for profit quickly drew Frenchmen into the interior, and the waters of the east coast of Canada were soon left behind. The maps and charts of Samuel de Champlain, one of France's greatest explorers and marine navigators, remained the basis of all maps of New France for many decades following his death in 1635, but he left the coast of Acadia as early as 1608.[1] Members of the Society of Jesus, who were the primary geographers of New France for many years after Champlain, likewise moved into the interior. Though producing work of such advanced detail and accuracy that it remained unsurpassed until the nineteenth century, as witnessed by the map of Lake Superior published with the Jesuit *Relation* of 1672, they too focused on the interior of the continent.[2] But during the final third of the seventeenth century, when fur trade expansion, the search for mineral wealth and the Western Sea, and the formal taking possession of new territory chiefly directed French state map-making initiatives into the interior, new hydrographic endeavour appeared at Quebec. Changing circumstances pulled New France into the orbit of new marine science.

First, in 1663 King Louis XIV revoked the charter of the One Hundred Associates, the lords-proprietor of New France, and the colony became a French royal province. Second, the colony came under the influence of Jean-Baptiste Colbert, who between 1665 and 1669 served as, in turn, controller general of finances, chief sponsor and organizer of the Royal Academy of Sciences, and minister of the Marine. French colonies, including New France, came under the administration of the navy, and Colbert desired to produce a group of colonial pilots and masters and to chart accurately the waters leading to New France. For several years colonial authorities employed local people who were not at the forefront of the new science to meet demands for more accurate geographical information. In 1666, for example, Martin Boutet, sieur de Saint-Martin, was charged by Intendant Jean Talon with examining and licensing surveyors, inspecting compasses, and teaching navigation to likely lads in the colony. But Boutet was also a Jesuit lay brother principally known as the cantor and choirmaster in the parish church.[3] Third, two scientific advances – changes in the theoretical principles of mechanical timekeeping developed by Christiaan Huyghens and a new theoretical discovery of a method to determine longitude based on the eclipses of Jupiter's satellites by Giovanni-Domenico Cassini – stimulated French efforts to solve the age-old question of finding longitude at sea. All three circumstantial developments drew New France into the orbit of new science, including hydrography.[4]

The Royal Academy of Sciences directed its attention towards New France at a very early date; indeed, its venture to Acadia in 1670 was an early example of a modern, overseas, scientific expedition. Jean Richer, one of the academy's *élèves astronomes*, was sent to Acadia to test a Huyghens marine clock and an improved method for finding longitude based on new tables of lunar distances prepared by Cassini, who had recently been put in charge of the Paris observatory, built in 1668.[5] Richer's expedition produced only modest results, for the marine pendulum clock failed soon after leaving France and could not be repaired, and no trace of the longitude test has been found. Richer did manage to fix the latitude of two points on the coast of present-day Maine and New Brunswick and report his observations in seconds of arc. This degree of accuracy was radically new and reflected his use of telescopically equipped instruments fitted with filar micrometers such as Abbé Jean Picard had devised a few years earlier, but the observations were of little use to navigators who lacked charts of similar precision on which to locate the points.[6]

Soon afterwards, members of the academy became caught up in a fierce debate concerning the earth's shape that involved the prestige of French science. In France the measurement of the earth was never a political, commercial, or military problem. It was and remained a purely scientific one and was treated as such by members of the Paris academy. The solution to the all-important question of finding longitude at sea, a practical problem of immense difficulty, awaited new technical developments and scientific theory, but despite encouragement from the state, French science was not organized to solve such a problem. New France was viewed within the greater context of this scientific debate about the shape of the earth, rather than as a means to solve a practical problem.

But like many blanket statements, this one needs qualification, for the next scientific expedition to New France did have a practical end in view, the construction of a hydrographic chart of the St Lawrence River. Jean Deshayes, who was sent to accomplish the task, had long been associated with the Paris academy and may have accompanied Richer to Acadia to test the measurement of lunar distances to find longitude at sea. In 1682, as "His Majesty's engineer for hydrography," Deshayes had been sent on an extended mission to the island of Gorée, off Cape Verde, Africa, and to Martinique and Guadeloupe in the West Indies to determine longitudes.[7] Three years later the new minister of the Marine selected him to make astronomical observations at Quebec and to construct a chart of the St Lawrence River. It is significant that the king's navy sponsored Deshayes's voyage to Quebec, rather than the Academy of Sciences.

Named royal hydrographer for the duration of his mission, Deshayes met with considerable success in New France. Following his arrival in August 1685, he accompanied Governor-General Denonville on a 350-mile voyage up the St Lawrence River to its source at the mouth of Lake Ontario. Though still suffering from his recent ocean voyage, Deshayes landed frequently on shore to observe latitudes of many places, which he did by sighting the Pole Star with a quadrant. This crude procedure allowed him to report his observations only to the nearest five minutes of arc. But following his return to Quebec, he observed a lunar eclipse on 11 December 1685, and later he established a meridian through Quebec, reported to seconds of arc. This precise determination of a longitude satisfied the demands of science for accuracy and met the colony's practical need for a meridian on which to construct surveys for land grants.

During the winter, Deshayes consulted local navigators, including Louis Jolliet, on ways and means to survey the St Lawrence. He travelled on snowshoes to establish baselines and lay down the initial triangles for his later survey of the river, and in May he departed in a bark provisioned for several months and manned by a crew of seven to spend the next six months working on the river.[8] His survey was incomplete when he departed for France in November, and plans were made for him to return the next year, but nothing occurred. Instead, Deshayes was succeeded as royal hydrographer in the colony by Jean-Baptiste-Louis Franquelin, the colony's first official map compiler, whose charts and maps are known chiefly for their colour and decoration, rather than their scientific accuracy.[9]

The fate of Deshayes's superb chart of the St Lawrence River, the precursor of all modern Canadian hydrographic charts, reveals the major problem arising from the absence of science in the daily affairs of either the navy or the colony. His survey lay hidden in the navy archives, along with other navigational and hydrographic observations, until the end of the century, for no office existed within the ministry to disseminate such information to those who could have used it. Locally printed charts of the Grand Banks and the coast of eastern North America had appeared in French ports, and pilots may have used them when sailing to Canada. But most navigators venturing across the North Atlantic employed Dutch charts, if they used charts at all, for they were cheap and readily available.[10] A French commercial chart-publishing industry never emerged in the face of Dutch competition and because of the organization of French science.

Colonial pilots and navigators worked to overcome hazards to local navigation, but their charts, such as one by Louis Jolliet compiled by Franquelin in 1685, were relatively crude even when based, as in Jolliet's case, on forty-six voyages made during the previous five years. In 1694, when on the Labrador coast, Jolliet observed the sun's height at noon with an astrolabe, a method employed by Champlain ninety years before.[11] It is unlikely that he knew how to use a telescopically equipped plane table such as Deshayes had used in 1686. Increased naval operations off North America during the 1690s gave rise to new demands for accurate charts and sailing directions based on improved scientific observations, and they probably lay behind the decision to publish Deshayes's chart of the St Lawrence River.

French naval officers knew next to nothing about North American waters and less about the dangerous voyage up the St Lawrence River to Quebec. They were forced to rely on ancient, inaccurate Dutch charts called *paescarts* to navigate the North Atlantic.[12] Many officers gathered sailing directions during their voyages to Newfoundland and Acadia, but most were incapable of making accurate astronomical observations. Organizational confusion also obscured the identification of problems, encouraged personal rivalries, led to uncoordinated and misdirected effort, and prevented successful resolution. The twofold attack on nautical problems by the Royal Academy of Sciences and the French navy achieved few practical results.

Changes within the navy's administration may have led to some improvement. In 1699 the first naval archives to hold ships' logs, navigation journals, sailing directions, and nautical charts was established. Much of this material had previously accumulated in ministers' offices, where it was of little use to anyone. Early in 1699 Jérôme de Pontchartrain, secretary of state for the Marine, responded to Jean Deshayes's request to publish his 1686 chart of the St Lawrence River by asking the Academy of Sciences to assess its accuracy. The academy replied quickly. In June Deshayes received royal permission to publish his chart whenever he judged appropriate, and the chart appeared early in 1702.[13] That same year he was appointed professor of hydrography at Quebec and sent to teach navigation and license pilots, but he died in 1706, too soon to have had any real impact on improving navigation. That year, after being unable to locate Cape Sable, at the southwest tip of Nova Scotia, one navy captain complained that, "once more, My Lord, the charts are worthless."[14]

Continual complaints about pilot ignorance and the inaccuracies of sailing charts led to calls for new surveys, but the War of the Spanish Succession and a crisis in navy finances halted attempts at improvement until after Louis XIV's death in 1715. That year Deshayes's chart was reprinted, amended with sailing directions compiled from navigation journals kept by Pierre Le Moyne d'Iberville during the 1690s. Afterwards, the dissemination of hydrographic information deteriorated. First, the duc d'Orléans, the Regent of France, and his advisers had more serious problems than hydrographic ignorance of North American waters to deal with. Second, Dutch commercial chart selling, though in decline, continued to prevent a French chart-printing industry from emerging without strong government support. But the growth and institutionalization of the Royal Academy of Sciences' power may have been the greatest obstacle.

The Paris academy's prestige and continuing evolution integrated it into the *ancien régime*. Its control of royal privileges for producing scientific instruments and publishing scientific work increased its authority over all scientific endeavour in France.[15] The *Connaissance des temps,* tables of lunar distances, which had appeared regularly since 1679, passed to the academy's control in 1702.[16] Deshayes had published his chart of the St Lawrence River only after receiving the academy's imprimatur and a royal privilege. Such control obstructed speedy dissemination of new information, gave the academy a monopoly over all scientific publication, and led science away from seeking solutions to practical problems. The continuing presence of serious theoretical questions concerning the measurement of the earth, which received the major attention of academicians, also contributed to the academy's delay of developments of an applied or practical nature. It enhanced the obsession of French map-makers with theoretical cartography, which was especially detrimental to applied hydrography. The controversy between Cassini and Sir Isaac Newton over the earth's oblate or prolate shape attracted almost all the geodetic attention of academicians and lasted to the mid-1740s.[17] Thus in 1735, while Pierre de Maupertuis and Charles-Marie de La Condamine were initiating their famous geodetic expeditions to Lapland and Peru to measure the degrees of latitude within the Arctic Circle and near the equator, the captain of a warship en route to Quebec complained that the Grand Banks were located too far east and the whole south coast of Newfoundland was defective on the Dutch chart he was using.[18]

New changes within the navy did not encourage nautical science. Although the department established a Dépôt des cartes, plans et journaux, a forerunner of what might be called an hydrographic office, at Paris in 1720, and the director was a naval officer with an engineer-geographer under his command, he had no clear mandate. He did as much or as little as he pleased to collect, collate, and disseminate hydrographic information to naval officers. During the next thirty years, the Dépôt des cartes, plans et journaux remained little more than a sinecure in the hands of a naval courtier and had little impact on the development of hydrography.[19] There is also little evidence to support an old claim that during the 1730s and 1740s the secretary of state for the Marine, the comte de Maurepas, stimulated the study of nautical science among naval officers.[20] The close relation between the government and the Royal Academy of Sciences prevented the former from seeing naval officers as more effective solvers of practical problems of nautical science than academicians. If anything, the opposite was true. It was *adjoint-chimiste* La Condamine's experience on board a navy warship that developed his interest in astronomy and navigation.[21] The policy of placing academician observers on board warships generally excluded naval officers from scientific investigation. When Maurepas sought intelligent young men and junior officers for further education, they came to his attention through the personal interest of a few captains and received additional instruction from civilians. Jesuit professors of hydrography, rather than naval officers or ships' masters, taught mathematics and navigation to young cadet officers, called *gardes de la Marine*.[22] The personal inclination of a few should not to be mistaken for the intellectual curiosity of a group.

Some improvements did occur between 1715 and 1750, but chiefly at a local or colonial level. Jean Deshayes's reissued chart of 1715 became the basic navigational tool of pilots sailing to Quebec during the remainder of the French regime, and his influence continued to be felt later on. His chart was the basis of the first English chart of the St Lawrence River, published in 1757 by Thomas Jefferys. In 1714 and 1716 Captain de Voutron, commanding *L'Afriquain*, made running surveys of the St Lawrence from Île d'Orléans to Kamouraska and in 1720 complained about errors in Deshayes's chart, but his proposal to survey systematically the waters of New France went nowhere.[23] In 1723 a naval officer placed navigation buoys in the St Lawrence River, and two years later a pilot and ship captain succeeded a local merchant as port captain (*capitaine de port*) at Quebec.[24] During the next sixteen years Richard Testu de La Richardière, aided by navy pilots left at Quebec over the winter, worked steadily to increase navigational security. They added to hydrographic knowledge and surveyed the river and gulf of St Lawrence as far as the Strait of Belle Isle and the south coast of Newfoundland. They also placed large range markers on masonry foundations on Île d'Orléans and cleared a one-thousand-foot strip through the woods to aid navigation through the area known as the Traverse. Surveying was extended into the Gulf of St Lawrence. Île Saint-Jean (Prince Edward Island), Baie des Chaleurs, and the Strait of Canso were all charted before 1741. La Richardière died that year, and succeeding port captains contributed little, but in 1744 the Crown appointed Joseph-Pierre de Bonnécamps (1707–90) to teach hydrography at Quebec.

Jesuit father Bonnécamps, unlike some of his predecessors, paid serious attention to his instructional and scientific duties, using the most advanced instruments of his day to teach young pilots. Colonial officials repeatedly on his behalf sought lodestones, pendulums, an observational telescope, and a quadrant with a three-foot radius mounted with a telescope rather than sights to support his work. He even possessed Jean Deshayes's seventy-year-old plane table and knew how to use it. But his complaint about the poor quality of the watch he used to determine longitudes suggests he did not possess a regulator or similar pendulum clock for measuring time.[25]

Recent changes in international relations and the serendipitous presence of Roland-Michel Barrin, marquis de La Galissonière, in New France and Paris inspired a surge of activity in hydrographic science and directed special attention toward Canada during the 1750s. The Treaty of Aix-la-Chapelle in 1748, the restoration of Louisbourg to France, the founding of Halifax by the British the following year, and the recall of the Marquis de La Galissonière from New France to become director of the Dépôt des cartes, plans et journaux, confidential adviser to the new minister of Marine, and member of the Anglo-French boundary commission appointed to settle territorial claims in North America brought home to the French government that, geodetically speaking, it did not know the location of the continent's eastern coastline.

Immediately following his appointment, La Galissonière began to transform the Dépôt des cartes into a modern hydrographic office, the first of its kind in the world. He organized an important scientific expedition, the first of three during his tenure, to carry out astronomical observations at Île Royale (Cape Breton Island). Significantly, the expedition was not entrusted to a member of the Paris academy but to a junior naval officer, Joseph de Chabert de Cogolin, who had just completed two years studying astronomy at Paris.[26] Chabert was sent to determine the latitudes and longitudes of Louisbourg, Canso, Sable Island, and Cape Race, Newfoundland. He was also to make observations of tidal currents and compass variations. He first made an extensive survey of the waters around Cape Breton Island and charted the Strait of Canso. During the winter he erected an observatory on the walls of Louisbourg, and despite the rigours of the climate, equipped only with a quadrant of three-foot radius and a pendulum clock, he successfully established the town's longitude. In the summer of 1751 he located Cape Sable and coasted the southeast shore of Nova Scotia. Before returning to France, he located Sable Island and then surveyed the south coast of Newfoundland from Cape Ray to Cape Race. True to his instructions not to waste time, Chabert reported his first observations to La Galissonière in the autumn of 1750. These were doubtlessly incorporated into the French memorandum delivered to the boundary commission.[27] International diplomacy and La Galissonière's personal interest inspired the renewal of hydrography in New France. Chabert's chart of Canada's east coast, for the first time based on astronomically

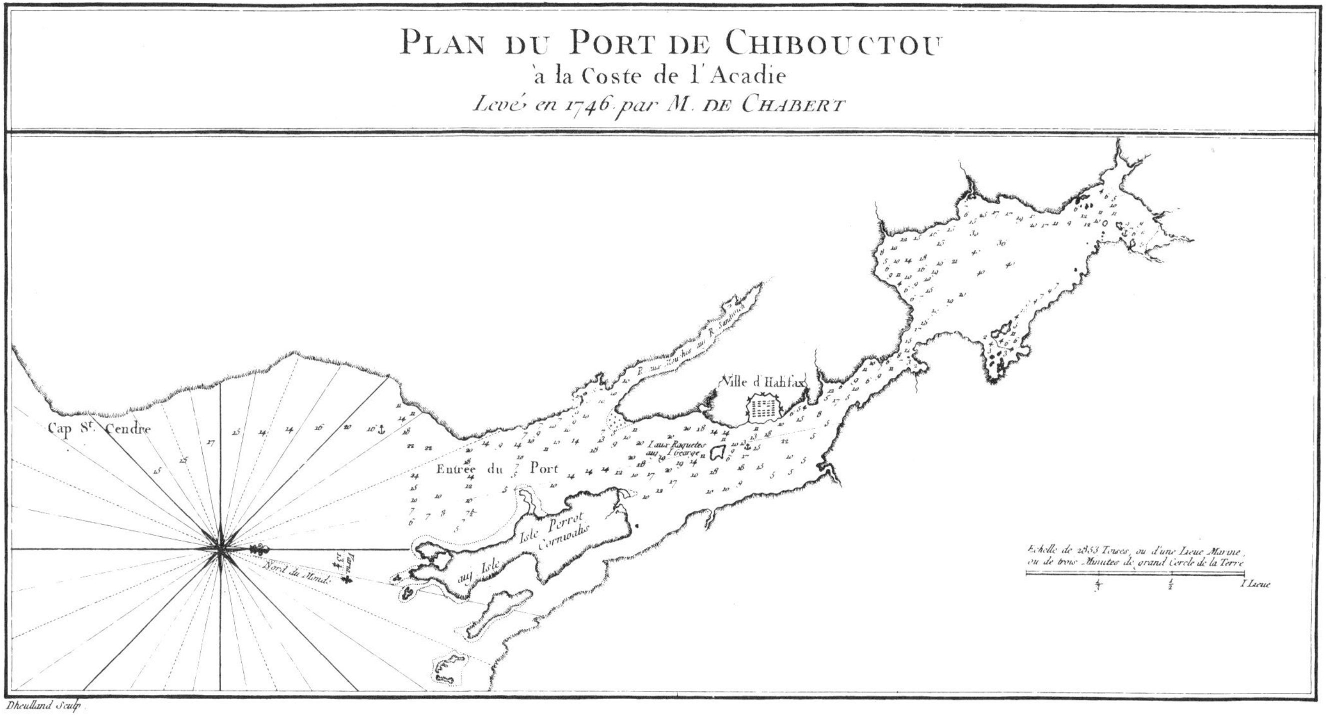


Plan du Port de Chibouctou (Halifax Harbour and Bedford Basin), by Joseph de Chabert de Cogolin, 1746. Drawn at age twenty-two while Chabert was serving as a *garde de la Marine*, or midshipman, in the frigate *Le Castor*, the chart shows the new English settlement of Halifax. (From Chabert de Cogolin, *Voyage fait par ordre du roi en 1750 et 1751*)

determined longitudes, shortened the coast of Nova Scotia by more than fifteen leagues and moved it nine degrees of longitude (about one thousand kilometres) closer to Europe than was shown on existing Dutch and English charts.

The second important development during La Galissonière's term as director of the navy's hydrographic office contributed greatly to practical navigation, for with his encouragement, the navy began to publish its own nautical charts and sailing directions. Joseph-Nicolas Bellin, the engineer-geographer employed at the Dépôt des cartes, plans et journaux for nearly three decades, had published a few charts prior to 1750, but the next year he produced a small, officially sponsored *Atlas maritime*. Two years later he issued a new edition of the sixty-year-old *Neptune françois* after the plates came into the navy's possession.[28] The reissued charts were out of date, but the *Atlas* and the *Neptune* broke the Royal Academy of Sciences' monopoly on official publication of scientific information. Finally, in 1756 Bellin issued the officially sponsored, two-volume *L'Hydrographie françoise*, including eighty charts of all the known coasts of the world, based on hydrographic surveys and sailing directions collected and collated over many decades in the Dépôt des cartes. This work gave French naval officers the most advanced printed aid to navigation in the world.

In addition, the marquis de la Galissonière persuaded the minister of the Marine to acquire the navy's own small observatory in Paris. It was located in the octagonal tower that today forms the entrance to the Hôtel de Cluny on the Rue du Sommerard.[29] He also arranged for Joseph-Nicolas de l'Isle to be attached to the Dépôt des cartes, plans et journaux as naval astronomer and for the department to purchase his vast collection of astronomical and geographical data, assembled from his own and his more famous brother's correspondence.[30]

Though further work at the Dépôt des cartes declined following La Galissonière's reappointment to sea duty on the eve of the Seven Years War, the office never ceased publishing hydrographic charts. Nicolas Bellin also responded to informed criticism. In 1761 he published a new chart of the St Lawrence River, amended and corrected on the basis of criticism from Canada. Three years later he issued charts of the east and west coats of Newfoundland.

The situation at Quebec improved during the final years of the French regime as a result of the work of Gabriel Pellegrin, but he fell a victim of his own probity and lack of influence. A former navy pilot who had accompanied La Richardière on his hydrographic surveys during the 1730s and early 1740s, he had returned to Quebec in 1751 to guide naval vessels in the St Lawrence River. He was named port lieutenant in 1752 and royal hydrographer succeeding Father Bonnécamps in 1757. Wartime duties kept him busy, but colonial and military authorities ignored his advice not to rely on the navigational hazards of the St Lawrence River for the defence of New France. Indeed, they paid no attention to the most knowledgeable person about the St Lawrence River in New France and ignored him completely during the siege of Quebec.[31]

That hydrography appeared in New France should be no surprise, for French scientists were at the forefront of the seventeenth-century scientific revolution. But the history of hydrography in New France, like Gabriel Pellegrin's career, nicely encapsulates the clash between theoretical and applied science and the interplay of character and circumstance that bedevilled its development. This was not to be the last time such conflicts occurred in the development of hydrography in Canada.

2

# *Alejandro Malaspina's Survey Operations on the Northwest Coast, 1791–1792*

ANDREW DAVID

Alejandro Malaspina was born in 1754 in the small town of Mulazzo, in the Ligurian region of Italy, the younger son of the Marquis Carlo Morelo Malaspina. Through his mother's family, he had an early connection with high Spanish authorities. He was first educated in the Clementine College in Rome; its library contained much of the literature of the Enlightenment, which the young Malaspina eagerly absorbed. Malaspina's father and uncle had hoped that the boy would become a priest, but he wished to pursue a naval career, to which they reluctantly agreed. At the time, Italy did not exist as a separate country, but since the Kingdom of the Two Sicilies was then under the rule of a son of the king of Spain, it was not unusual for Italians with military or naval ambitions to enter the Spanish service. First, however, Malaspina served a cadetship in the Order of Malta, which had a minuscule navy, thus strengthening his credentials in Spain. Together with family influence, this experience was enough to secure him a place in the midshipmen's college in Cádiz on 18 November 1774.

Malaspina's early training in the Order of Malta stood him in good stead, for in January 1775 he was promoted to *alférez de fragata*, the most junior commissioned rank in the Spanish navy, and appointed to the frigate *Santa Teresa*. Further promotions followed rapidly. As a *teniente de fragata* in the *San Julián*, he took part in the Moonlight Battle off Cape St Vincent,[1] in which a British squadron under Admiral Rodney defeated the Spanish fleet under Juan de Lángara. Promoted to *capitán de fragata* on 21 December 1782, Malaspina was given command of the frigate *Asunción* and sent to Manila to inform the authorities there of the end of the war with Britain. On his return to Spain, he joined a group of naval officers studying astronomy at the observatory at Cádiz, at the request of Vicente Tofiño de San Miguel. Among these officers were Dionisio Alcalá Galiano,[2] José de Espinosa y Tello, and Juan Vernacci y Retamal, all of whom later served under Malaspina on the Northwest Coast of North Amercia.

On 15 September 1786 he sailed once again for Manila in the *Astrea*, on a commercial voyage sponsored by the Real Compañía de Filipinas. One of the passengers was the supercargo Manuel Agote, who wrote to the company from Lima reporting that Malaspina's "meticulous work in the calculation of longitudes promises that his future voyages will be carried out brilliantly."[3]

On his return to Spain, Malaspina and his friend José Bustamante de Guerra wrote to Navy Minister Antonio Valdés y Bazán on 10 September 1788, proposing a scientific and political voyage around the world "following earnestly in the wake of Cook and La Pérouse."[4] Valdés replied on 14 October 1788 that Malaspina's project had merited the king's approval. In consequence, work was put in hand at once to build two identical corvettes in Cádiz of 306 *toneledas*,[5] with an overall length of 120 *pies* (110 feet), to be named *Descubierta* (Discovery) and *Atrevida* (Daring). Malaspina was appointed to command *Descubierta* and, as the senior officer, the expedition as well, with Bustamante in command of the *Atrevida*.

Malaspina was given a free hand in the choice of his officers. These included Galiano and Vernacci, who had both, as noted above, served under Tofiño and were thus competent hydrographic surveyors. Galiano had also served under Antonio de Córdoba y Lazo in the *Santa María de la Cabeza* during a surveying voyage in 1785–86 to the Strait of Magellan, as had Ciriaco Cevallos y Bustillo, who joined the expedition in Acapulco with Espinosa. As a result, Galiano was a more-than-competent astronomer as well as a talented surveyor. A key appointment was that of Felipe Bauzá y Cañas as cartographer, whose task was to draw the surveys. However, he also carried out surveys and was, in addition, a competent topographical artist. Most of the coastal views executed on the voyage were drawn by him. Another excellent choice was Cayetano Valdés y Flores, nephew of the navy minister. Among the supernumeraries when the expedition left Cádiz were Colonel Antonio Pineda as director of natural history and the botanist Luis Neé. Tadeo Haenke, a naturalist from Prague, joined the expedition in Valparaiso after many adventures. It should thus be clear that when Malaspina arrived on the Northwest Coast, he brought with him a talented collection of officers and scientists, whose combined expertise compared most favourably with both James Cook and the comte de La Pérouse.

A great deal of effort was exerted to obtain the latest surveying instruments. As the planning progressed, it soon became apparent that the Royal Observatory in Cádiz did not hold sufficient instruments to equip the two corvettes. Steps were therefore taken to purchase a considerable number of instruments in London, where the principal instrument makers at the time operated. Thus on 31 October 1788 Malaspina wrote to Valdés enclosing a list of instruments required for the voyage and asking for them to be obtained in London;[6] it included sextants, artificial horizons, and an Arnold chronometer.[7] Instructions were sent to Juan Jacinto Magallanes, a Portuguese astronomer living in London, asking him to approach Alexander Dalrymple, hydrographer to the East India Company, for his help in acquiring these instruments, particularly watches and chronometers by John Arnold.[8] Malaspina followed up this request by writing personally to Dalrymple on 17 January 1789.[9] On 19 January Valdés also wrote to Fernán Núñez in Paris soliciting instruments for the expedition.[10] Once again Malaspina followed up this inquiry by writing the following day to the French astronomer Joseph-Jérôme Lalande.[11] An astronomical or regulator clock, an astronomical quadrant by Ramsden, two theodolites, and a 100-foot surveyor's chain, probably the well-known Gunter's chain, were also obtained from London and issued to the *Descubierta* by the Royal Observatory, which as well provided the corvette with a 100-foot chain from its own collection.[12] A portable observatory, similar to that used on Cook's three voyages was also obtained. After the expedition sailed, the Spanish naval officer José Mendoza y Ríos, a noted mathematician and astronomer resident in London, obtained additional instruments, including a special pendulum for measuring the force of gravity at different locations.[13]

Perhaps the most important instruments held on board the two corvettes were the chronometers. In April 1791, before the expedition departed from Acapulco for the Northwest Coast, Malaspina listed the chronometers: Arnold numbers 61, 71, and 72 on board the *Descubierta* and Berthoud number 10 and pocket chronometer Arnold number 351 on board the *Atrevida*, augmented by Bustamante's personal pocket chronometer, Arnold number 105.[14] Arnold number 351 had been brought to Acapulco by Cevallos and Espinosa, together with the special pendulum for the gravity experiments. They had also brought with them Arnold pocket chronometer 344 for the *Descubierta*, but Malaspina issued this instrument to Galiano, who was being sent to Mexico City to gather navigational information and carry out astronomical observations there during the expedition's cruise to the Northwest Coast.

A considerable library of books pertaining to navigation and hydrography was also ordered for the expedition from London. These included a complete set of Cook's *Voyages*, a complete set of Dalrymple's charts and sailing directions, Thomas Jefferys's *American Atlas* of 1776, *Mayer's Lunar Tables Improved by Mr. Charles Mason*, the British *Nautical Almanac* for the years 1789, 1790, and 1791 (three copies of each),[15] and copies of the French *Connaissance des temps*. In April 1789 Valdés sent Malaspina a summary of La Pérouse's voyage from Manila to Kamchatka, which he had obtained from Paris.[16] A reference made by Malaspina in his journal to holding Nathaniel Portlock's and George Dixon's surveys[17] indicates that he had their *Voyages* on board as well, while John Meares's *Voyages* was obtained after the expedition sailed and forwarded to Acapulco in time for it to be received on the expedition's second visit to the port.[18] References to William Dampier's and George Anson's *Voyages* make it clear that Malaspina also held other voyages on board, possibly in their Spanish editions.

A major part of his mission was to carry out detailed coastal and harbour surveys, each of which required different techniques. As a result of having served under Vicente Tofiño during the latter's extensive survey of the coasts of Spain, Malaspina adopted survey methods mainly based on those of his mentor. For coastal surveys he employed a technique known as a running survey, in which a series of "ship stations" were established, the distance between them being computed from the time taken for the corvette to sail between the stations and the speed of the corvette as measured by log-line, a procedure that Malaspina referred to as measuring bases. The bearing between each ship station was established from the courses steered by magnetic compass with due allowance for leeway and currents. At each ship station, compass bearings were taken to prominent objects on shore and the angles between them measured by horizontal sextant angles, enabling their positions to be established relative to the *Descubierta,* a procedure that Malaspina sometimes referred to as observing triangles. Observations to obtain magnetic variation were taken each day when possible, so that magnetic bearings could be converted to true bearings. To achieve the best possible accuracy in his running surveys, Malaspina placed the knots on his log-line 50⅔ English feet apart, which, in conjunction with a thirty-second sand glass, represents a nautical mile of 6,080 feet, close to the present international value of 1,852 metres. His use of English feet rather than Burgos feet, which were slightly shorter,[19] suggests that Malaspina had consulted the comments made by Constantine John Phipps, who experimented with various log-lines during his voyage to Spitzbergen in 1773; a copy of his account was held on board the *Descubierta*.[20]

Bases were also sometimes measured at sea by observing the angles of elevation between the waterline and the top of the masthead of the other corvette, from which the distance between the two corvettes could be calculated by simple trigonometry, a method also described by Phipps.[21] Horizontal sextant angles were observed simultaneously from each corvette between the mainmast of the other corvette and selected objects on shore to fix them in relation to the base. When bases were measured by this method, it was essential to exchange data as soon as possible to ensure that any points that were observed from both corvettes were correctly identified. On a number of occasions, Bauzá was sent over to the *Atrevida* for this purpose. However, Malaspina found this method to be unreliable when the corvettes were under way, probably because of the difficulty in achieving simultaneous observations. To establish the geographical framework for these running surveys, meridian altitudes of the sun were observed, when possible, to obtain latitude, while longitude was obtained by lunar distances or by chronometers. In the latter case, Malaspina had first to obtain apparent time by observing the altitude of the sun when it was some distance from the meridian. Then, by knowing the corvette's latitude and the sun's declination, it was possible to calculate the sun's hour angle at the time of the observation. If the sun's altitude had been observed before noon, as was usually the case, the hour angle was then subtracted from twenty-four hours to obtain apparent time. This was then converted to local mean time by applying the equation of time taken from the British *Nautical Almanac* or the French *Connaissance des temps*. The difference between local mean time and mean time at Cádiz, as given by the corvette's chronometers, would provide her longitude west of Cádiz expressed as time. Malaspina referred to this operation in his journal as observing hour angles. At dusk he either lay to or stood off and on, with the intention of having the final points fixed the previous evening visible at dawn, so as to enable him to continue his survey where he had left off the night before. After the positions of various points on land had been established, the intervening coastline would be sketched in by eye. Soundings were taken whenever possible, and the type of bottom recorded. When no bottom was obtained, the length of the lead-line was noted.

In harbour surveys, bases were measured by steel chains, graduated in English feet, from which a system of triangulation was then observed by theodolite. Early in the voyage, because of unsuitable terrain, Malaspina first attempted to measure a base by masthead heights, but for some unknown reason, this proved unsatisfactory, and Bauzá was sent ashore to measure additionally a short base by chain. Soundings were usually fixed by means of compass bearings taken from the boats. Occasionally, however, two theodolites were set up on shore, and angles were taken to the sounding boat, the timing probably controlled by flags or pistol shots, using a method devised by Tofiño. When possible, the surveyors would fix the position of the coastline by walking along it.

At each place where the expedition stopped for more than a few days, Malaspina landed the expedition's portable observatory, the astronomical clock, the astronomical quadrant, various telescopes, and several chronometers, in order to establish the observatory's position and to rate his chronometers. The latitude of the observatory was obtained by means of meridian altitudes of the sun and various stars, while longitude was obtained by a variety of methods, namely, by lunar distances, chronometer, the occultations of the satellites of Jupiter, or the occultation of stars by the moon. At first, Malaspina's longitudes were given from Cádiz, but later they were calculated from the last place where he was confident that he had obtained its longitude with sufficient accuracy. The rates of his chronometers were obtained by means of equal or corresponding altitudes. Comparisons between the astronomical clock and chronometers on shore and the chronometers left on board were again carried out by means of pistol shots, the time of the flash rather than the sound of the shot being noted.

Most of the astronomical observations taken on shore were carried out by Galiano, Vernacci, and Juan Gutiérrez de la Concha, whom Malaspina regarded as the expedition's astronomers. As well as making the necessary observations for rating the chronometers and establishing the geographical positions of the various observatories as part of the surveying program, earlier in the voyage they had also carried out observations to establish the fundamental positions of stars in the southern heaven.

When Espinosa and Cevallos joined the expedition on its first visit to Acapulco in March 1791, they brought with them instructions dated 22 December 1790, from Navy Minister Valdés, ordering Malaspina to carry out observations for gravity with the specially designed pendulum, which these two officers had brought with them.[22] This instrument, which was probably based on one used by Israel Lyons in 1773 in Spitzbergen during Phipps's voyage,[23] had been obtained in London by Mendoza y Ríos.[24]

As is well known, the earth is not a perfect sphere, being flattened at the poles because of the pull of gravity and the rotation of the earth as it cooled. The purpose of these observations, as explained to Malaspina by Valdés, was to obtain information about the true figure of the earth. In particular, observations were required in 45° S to determine whether the Southern Hemisphere was flatter than the Northern Hemisphere in the same latitude. Similar observations were being conducted in the Northern Hemisphere by the French. Cevallos and Espinosa had made a number of observations with the pendulum before sailing for Acapulco, probably in Cádiz. Three observers were required to carry out the observations. Two of them counted the number of oscillations in an hour, while the third recorded the time by chronometer. During the voyage, observations were made at eight locations in the Northern Hemisphere and at seven in the Southern Hemisphere. After the voyage, the results were computed in 1807 by Brigadier de la Real Armada Don Gabriel Ciscar, who obtained a value of compression of 1/321; this result was published two years later by Espinosa.[25] It compares reasonably well with the value of 1/298·257223563,which was adopted for the World Geodetic System in 1984.

Malaspina was aware of a more time-consuming method of determining the figure of the earth which had been devised by the French astronomer Giovanni-Domenico Cassini. In this method an accurate base was measured on land and extended by triangulation to two points on the earth's surface in a north-south line and approximately one degree apart. From the difference in latitude between the two points, obtained astronomically by meridian altitudes, and the distance between them, obtained through triangulation, the length of a degree of the meridian at the mid-latitude of the observations would be obtained. Since the length of a degree of latitude is dependent on the radius of curvature of the earth along the meridian, two or more observations in widely different latitudes would enable the compression to be calculated. This method was referred to by the Spaniards as *medida del grado terrestre* (measurement of the terrestrial degree), but British scientists always referred to it as the measurement of an arc of the meridian, since it was highly unlikely that the two extremities would be exactly one degree apart. In 1735 two expeditions had been mounted by the French, one to Lapland and the other close to the equator in the vicinity of Quito, in present-day Ecuador. The later expedition, which took eight years to complete, was led by Pierre Bouguer and Charles-Marie de La Condamine, with two Spanish naval officers, Antonio de Ulloa y de la Torre-Guiral and Jorge Juan, ostensibly as observers, all four were mentioned by

Malaspina in his journal.[26] From an earlier measurement of an arc of the meridian by Cassini near Paris, the length of a sea mile (one-sixtieth of a degree of latitude) at that latitude was determined accurately. This value was used by Malaspina to determine the correct distance apart of knots on his log-line.[27]

On his arrival at Acapulco on 27 March 1791, he learnt that Espinosa and Cevallos had already arrived at this port on 26 February, bringing with them the special pendulum and pocket chronometers Arnold 344 and 351. Since the *Descubierta* was not expected for some weeks, they joined the *Atrevida,* which had been been sent ahead and was waiting for them. The following day Bustamante sailed for San Blas, where he was to consult with Juan Francisco de la Bodega y Quadra and to obtain a large launch similar to one that had earlier been reconstructed for the *Descubierta* in Guayaquil. Meanwhile, in Acapulco, Malaspina received instructions from Navy Minister Valdés that his first priority was to sail north to search, in the vicinity of 60° N, for the western entrance to the strait through which Lorenzo Ferrer Maldonado was reported to have entered the Pacific in 1588.[28] First, however, Malaspina paid a visit to Mexico City to consult with the governor, Conde de Revillagigedo, and the Galician mariner Francisco Antonio Mourelle de la Rúa, who had extensive knowledge of navigation along the Northwest Coast of North America. While he was in the capital, Malaspina examined some astronomical instruments belonging to the Royal Observatory at Cádiz, which included an astronomical or regulator clock by Ellicott[29] and an astronomical quadrant by Adams.[30] These instruments were used by Galiano in Mexico City during the expedition's cruise in northern waters.

Bustamante rejoined Malaspina on 20 April, enabling the two corvettes to sail from Acapulco in company on 1 May to search for Ferrer Maldonado's strait. Malaspina first stood well offshore to avoid winds from the north and to obtain those from the northeast at a better angle, intending to make the coast in the vicinity of Cook's Cross Sound and Cape Fairweather. From here he planned to survey the coast as far as Kodiak Island, which Cook had not examined in detail, paying particular attention to Bherings Bay on Cook's chart, behind which the English captain had noted a stretch of low coast with no backing hills, leaving him in some doubt that there might be a body of water to the north.[31] To assist him in his search, Malaspina constructed, on his way north, a large plotting sheet extending from the northern extremity of Bucareli Bay[32] to Kodiak Island, based on the British surveys of Cook and Portlock[33] and the Spanish surveys of Bodega, Ignacio de Arteaga, Salvador Fidalgo, Estéban José Martínez, and López de Haro and using as a principal point of reference the longitude determined by Cook for Cape Edgecumbe, which Malaspina called Cabo de Engaño. It was from this cape that he intended to begin his running survey.

On 23 June land was sighted, which proved to be the stretch of coast between Cape Edgecumbe and islands to the north of Cape St Bartholomé.[34] Mount Edgecumbe was soon identified, Malaspina commenting that the stupendous height of this mountain caused it stand out from the rest, making it a very useful landmark. To help others identify it, Bauzá included a view of the mountain in one of a number of *vistas*, or coastal views, he drew along this coast.[35] Its height was duly measured but left blank in Malaspina's journal. The snow-covered peak of Mount Fairweather was sighted next. As Malaspina followed the coast to the west, he imagined several times that he had sighted inlets which might be Cook's Bherings Bay. Eventually, however, he decided it would be best to examine the coast in more detail with the launches while the corvettes wooded and watered in Port Mulgrave, in present-day Yakutat Bay. In clear weather it was possible to observe several hour angles and measure some bases to progress the survey. This work and the clear weather finally dispelled any idea of a passage or even an inlet along this stretch of coast. Finally, at six o'clock in the morning of 28 June, Malaspina rounded Ocean Cape and entered Yakutat Bay, guided by Dixon's chart.[36]

As he entered the bay, he could see at its farthest point an inlet whose entrance and course seemed to tally with Ferrer Maldonado's description. After consulting with Bustamante, Malaspina therefore set a course towards the inlet. A pleasant east-southeast breeze and clear skies enabled him to observe several sets of hour angles and promised the possibility of observations for latitude at noon. However, as he approached the inlet, the *Atrevida* reported that she had been unable to obtain bottom with eighty fathoms of line, thus making anchoring impossible. Malaspina therefore decided it would be more

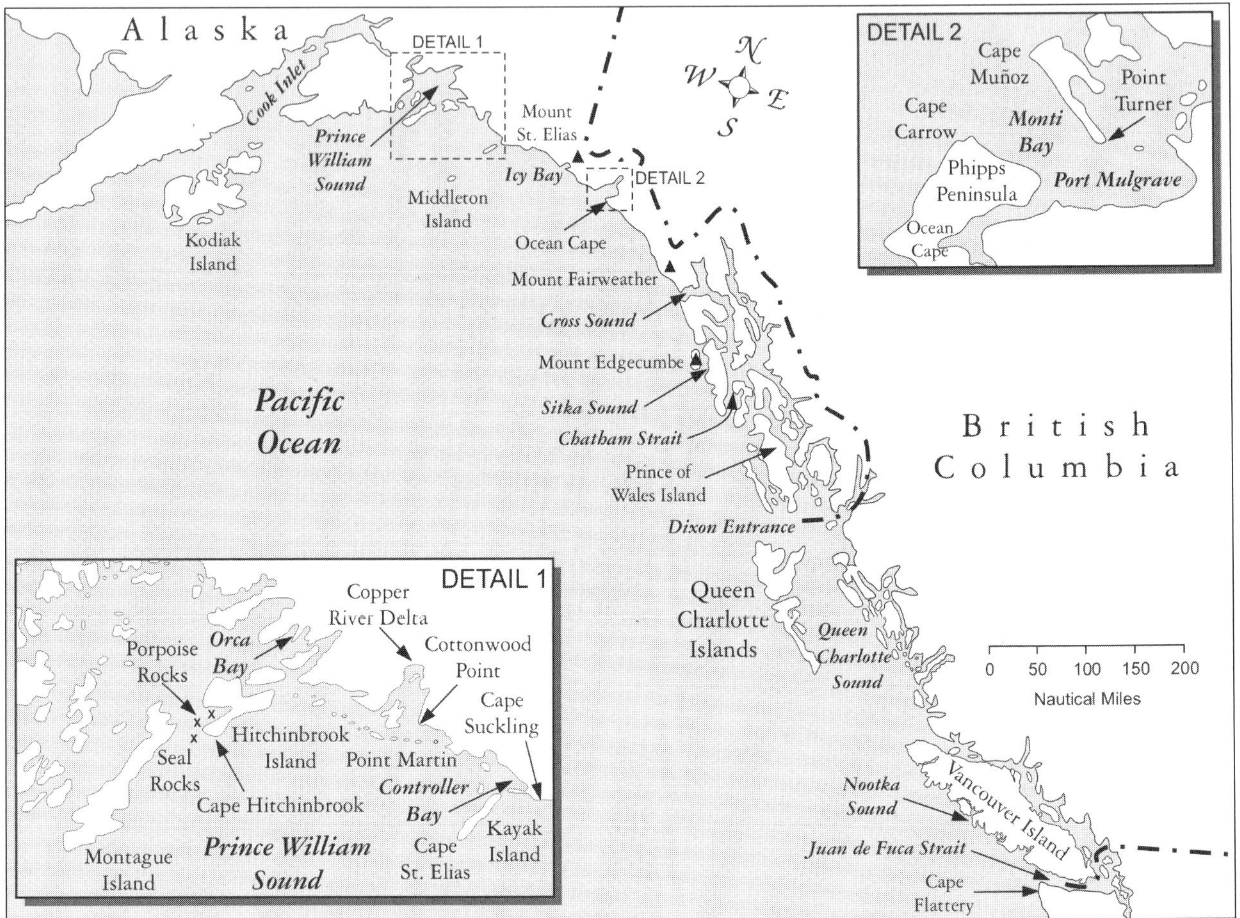


The Northwest Coast of North America (Map: Chris Johnson)

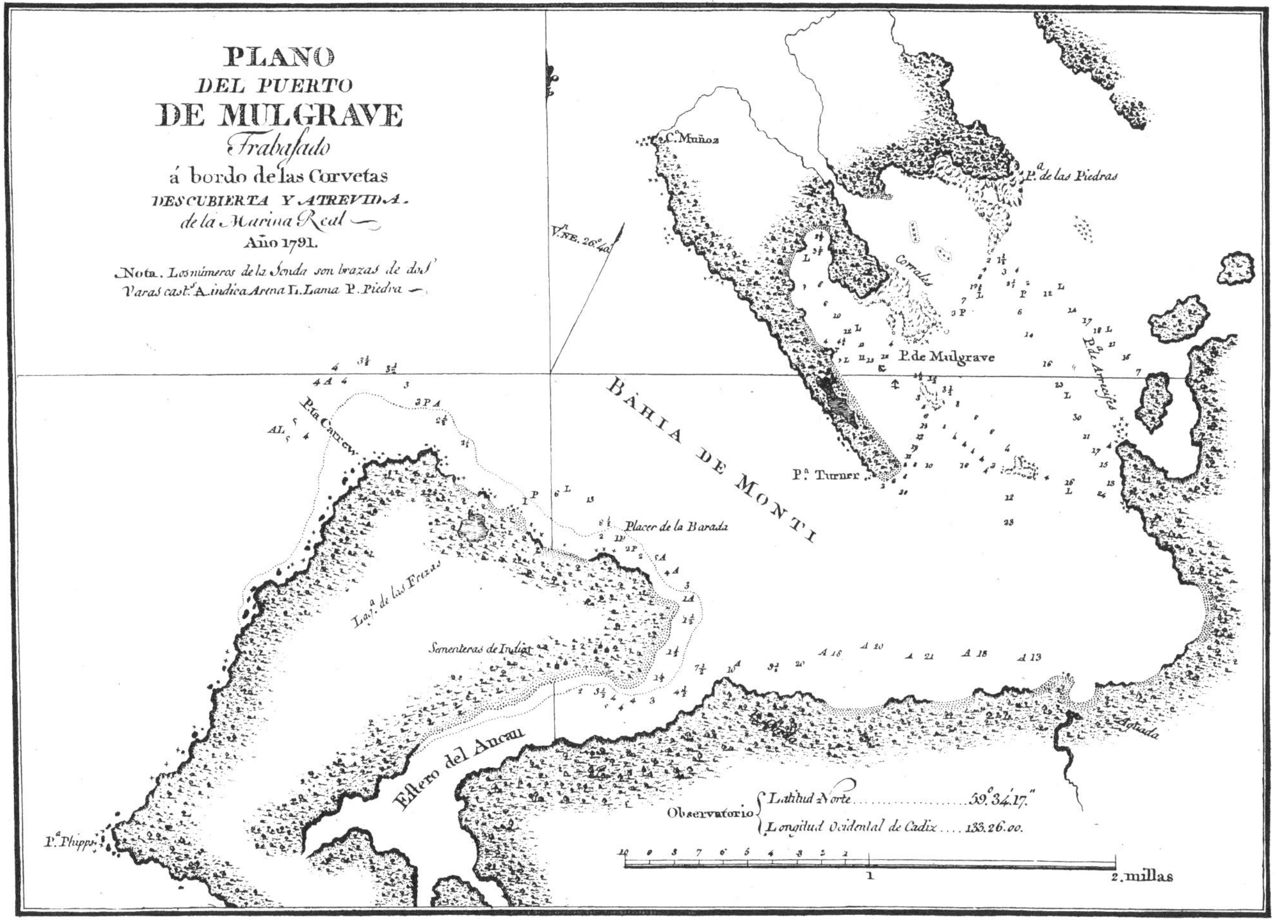


Plano del Puerto de Mulgrave. (From *Relación del viage*)

prudent to make for Port Mulgrave and anchor in the harbour already surveyed by Dixon and from there to undertake an examination of the inlet by launch. At seven o'clock that evening, the corvettes rounded Turner Point and, entering Port Mulgrave, dropped anchor in twelve fathoms opposite the Tlingit settlement there.

A suitable site for the portable observatory was found on Turner Point, which was covered by the corvettes' guns and, in addition, was some distance from the Tlingit village. It was also completely free from any hiding places in trees or any other shelter. The survey of the harbour began with Bauzá measuring a base near the observatory, taking bearings from both ends of the base to as many suitable objects as the somewhat overcast conditions would allow. Observations were taken at the observatory with the Ramsden astronomical quadrant for latitude and longitude and to rate the chronometers, while observations to obtain magnetic variation were made by theodolite. The pendulum was also set up near the observatory, and observations made to obtain the force of gravity. When the sky cleared, Mount St Elias bore N 38°59′ W by magnetic compass, which, by applying variation of 32°49′ E, gave its true bearing as 318°12′. Its height was also obtained from observations taken by theodolite from one end of the base line. The survey of the harbour was next begun, with Bauzá taking soundings in the entrance channel in the *Descubierta*'s fully armed pinnace, while Díaz Juan Maqueda, a junior officer, checked the soundings inside the harbour in the *Atrevida*'s pinnace, also fully armed.

Leaving Bustamante in charge of the two corvettes, Malaspina set off on 2 July with the two launches to examine the inlet at the north end of Yakutat Bay. Bauzá accompanied this expedition, landing at a convenient place to measure a base and take some bearings with a theodolite. A short distance beyond the narrow entrance to the inlet, further progress was blocked by permanent ice. Observations were taken for latitude and longitude, while several azimuth observations by theodolite were also obtained. From two further locations, Bauzá was able to obtain additional bearings of Mount St Elias. Forced to turn back, Malaspina expressed his disillusionment by naming the inlet Desengaño, now Disenchantment Bay.

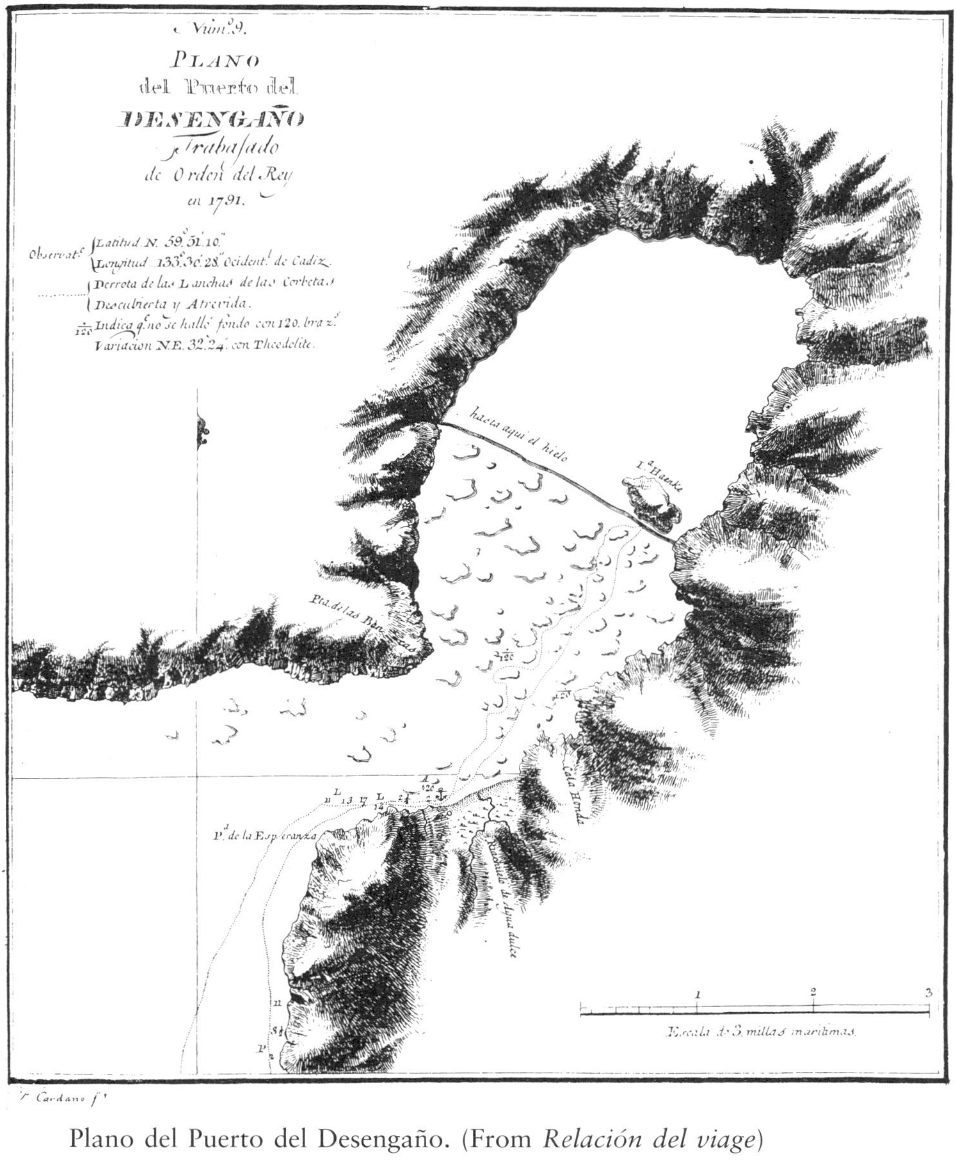


Plano del Puerto del Desengaño. (From *Relación del viage*)

On his return to Port Mulgrave after an absence of three days, Malaspina found that the Tlingit had become somewhat aggressive. Their actions culminated in an attempt to steal the Ramsden astronomical quadrant as it was being brought down to the beach to be taken on board after further observations to rate the chronometers. An ugly confrontation took place, which was only averted when a blank round was fired from one of the *Atrevida*'s guns. Malaspina was thankful to make his departure on 6 July without bloodshed.

He gives no details in his journal of the astronomical observations he took in Port Mulgrave. Fortunately, these were published by Espinosa in 1805 in a twenty-page pamphlet[37] and again four years later in a much more comprehensive work.[38] They show that observations for latitude were made on 30 June by sextant and artificial horizon and the following day by astronomical quadrant, giving a value of 59°34′20″ N. The calculation of longitude was far more complicated. First, a longitude of 40°19′52″ west of Acapulco was obtained by the expedition's four chronometers. Since that city was thought to be 93°42′50″ west of Cádiz, this figure gave the longitude of the observatory as 134°02′42″ west of Cádiz. But these values were then adjusted to take into account various lunar distances observed on the passage north, with the result that the mean longitude obtained by the four chronometers was adjusted to 133°18′00″ west of Cádiz. On 24 June a series of 200 lunar distances were observed which, carried forward by chronometer, gave a longitude of 133°27′22″ west of Cádiz, and a series of 334 observed on 23 July in sight of Mount St Elias, carried back to Port Mulgrave, gave a value of 133°27′15″ west of Cádiz. The mean of these three values gave the true longitude of Port Mulgrave, according to Espinosa, as 133°24′12″ west of Cádiz, or, assuming that San Fernando Real Observatorio is 6°12′16″ west of Greenwich, its longitude west of that meridian was 139°46′28″, which compares favourably with its present accepted longitude of 139°46′·9 W. Espinosa gave the height of Mount St Elias above sea level as 2,793 *toesas*[39] and its position as 60°17′35″ N and 134°33′10″ west of Cádiz,[40] or 140°45′26″ west of Greenwich, compared with its present accepted longitude of 140°56′ W. Malaspina's surveys of Port Mulgrave and Disenchantment Bay were published in 1802 as "Plano del

Puerto de Mulgrave trabajado á bordo de las corvetas *Descubierta* y *Atrevida* dė la Marina Real año 1791" and "Plano del Puerto de Desengaño trabajado de orden del Rey en 1791."[41]

On his departure from Yakutat Bay, Malaspina continued his running survey towards Kodiak Island. By now he had not even the slightest hope of finding Ferrer Maldonado's strait, but he felt that he had to continue his search for it to fulfill his instructions from Navy Minister Valdés. He was also aware that this stretch of coast had not been examined in detail either by various Spanish navigators or by Cook, Portlock, or Dixon. At first a moderate northwesterly or westerly wind brought with it good visibility, so Malaspina should have sighted the largest glacier in Alaska, now named after him, which nearly reaches the sea behind Sitkagi Bluffs and is situated approximately ten to twenty miles west of Point Manby, the western entrance point of Yakutat Bay. The wind, however, soon backed to the southwest and south and finally to the east, accompanied by poor visibility, making precise surveying impossible. He therefore decided to make for Prince William Sound in the hope of finding a suitable anchorage there until conditions improved. However, he took the opportunity to search en route for a shoal said to have been discovered in 1779 by the Spanish corvettes *Princesa* and *Favorita*.[42] But the search was discontinued when the wind backed to east-northeast, bringing with it improved visibility.

Malaspina decided instead to make for Cook's Cape Suckling, as indicated on Cook's chart, where he began his survey again, the *Atrevida* being responsible for obtaining soundings. When a moderate easterly breeze set in, he considered attempting to pass between Kayak Island and the cape in order to survey Controller Bay, but decided against it and continued his survey to the west. Off Montagu Island, José Sánchez was sent inshore to look for a suitable anchorage. His report, however, appears to have been unsatisfactory since Malaspina decided to search instead for some offshore islets named Hijosa after Francisco Hijosa, the commissary at the Naval Department of San Blas, by the Spanish navigators who had sighted them; as Malaspina pointed out, neither Cook nor Dixon had included them on their charts. He realized that Cook had not passed close enough to these islands,[43] but he was puzzled that Dixon had failed to sight them since his track, in his first year in these waters, took him through the area.[44] On 11 July Malaspina sighted a low island, which he rightly assumed to be Hijosa, although its position in both latitude and longitude was very different and it was a single island. He remained in the vicinity of the island for five days, enabling him to fix its position in 59°26′ N, 6°37′ west of Port Mulgrave. He then made an uncharacteristic decision, naming it Galiano Island after his absent astronomer, in spite of an earlier decision that he would respect the names given by first discoverers. Malaspina's name was not accepted for Spanish charts, however, and even Hijosa was superseded by Middleton Island when Vancouver's atlas was published.[45]

Malaspina now set a course for Cape St Elias, the southern extremity of Kayak Island, with the intention once again of attempting the passage between the island and Cape Suckling. On closer approach to Kayak Island, he became convinced, wrongly, that it was not an island but a peninsula joined to the mainland by a low thickly forested spit. He was, however, able to fix a number of features at the back of Controller Bay. He therefore altered course to the east towards Port Mulgrave, but poor visibility forced him to remain in the vicinity of Cape Suckling for several days until a Southwest wind cleared the sky, enabling him to continue his survey. He found no sign of an opening as far as Punta Verde,[46] not far from the foot of Mount St Elias. When the skies cleared, no opportunity was missed to obtain longitude by chronometers and lunar distances, while bases were measured and soundings taken whenever possible. Along this stretch of coast, Malaspina noted a comparatively shallow bay, which he named Estremadura. This was in fact Icy Bay, the name adopted later from its description in Vancouver's journal. The bay is much larger than Malaspina thought. Off the bay the wind dropped, forcing him to anchor in full view of Mount St Elias, to prevent the corvettes being set too far inshore. At dawn on 26 July, when a moderate breeze set in from the first quadrant, Malaspina got under way again and set a course for Cape Fairweather, which he reached at noon on the 28th. As he passed Mount Fairweather, he calculated its height to be 5,368·3 *varas*,[47] or 4,488 metres, slightly below the figure of 4,670 metres accepted today. From here Malaspina kept the coast in sight, as far as possible, mainly checking on the work of his predecessors rather than attempting a detailed survey. He had hoped to make a detailed examination of the entrance to Port Banks,[48] which Dixon had visited, but poor visibility and a strong easterly current thwarted his plan. At one stage, visibility was so bad that the corvettes had to keep in touch by firing their guns. On 3 August the northwest tip of the Queen Charlotte Islands was sighted, but the weather was so gloomy and threatening that Malaspina was unable to check its position. Two days later the corvettes were struck by a severe storm, which he considered worse than any they had experienced since leaving Spain. On 11 August, the storm having by now abated, they were off Cape Cook, on the northwest coast of Vancouver Island, and two days later the two corvettes were safely at anchor in Friendly Cove.

At the time of Malaspina's visit, there was a substantial Spanish presence in Friendly Cove under Teniente de Navío Don Manuel Saavedra, commanding officer of the frigate *Concepción*, with numerous buildings, including a hospital, and a vegetable garden. Once the usual formalities were over, the portable observatory, the astronomical instruments, and the special pendulum were landed and set up in the southwest corner of the cove, not far from some of the buildings of the Spanish establishment. In pleasant afternoon weather, Bauzá began taking bearings to many distant objects from the forecastle of the *Descubierta*, including the 1,481-metre-high Tahsis Peak, now Conuma Peak, a dramatic cone-shaped mountain that dominates the view up Tlupana Inlet, inside Nootka Sound, and was thus a very useful mark for the intended survey of the harbour, which began with the measurement of a base on shore.

For various reasons, Malaspina found that relations with the Natives had become somewhat strained. When Cayetano Valdés and Bauzá called at the village of the principal chief, Maquinna, during the survey, they found it deserted and its inhabitants hiding in the nearby forest; hardly any of them could be persuaded to approach the visitors. In consequence, much of Malaspina's personal endeavours during his stay in Nootka were directed towards improving relations with the Natives. He also realized that the Spanish were still unaware of how far the numerous internal channels extended inland and whether they communicated with the harbours further south or even with the Strait of Juan de Fuca. He was, however, aware that the American fur trader John Kendrick had made his way by the inner channels to Esperanza Inlet, thus establishing the insularity of Nootka Island. He also knew that some survey work had been carried out by a former commandant, Francisco de Eliza, but more clearly needed to be done.

In consequence, the expedition's two launches were made ready under the command of Espinosa and Cevallos, fully armed and with rations for nine days. They were equipped with an excellent Stancliffe sextant belonging to Espinosa, a Ramsden sextant, pocket chronometer 351, and a theodolite. Their instructions were that they were to commence their survey at Resolution Cove, Cook's anchorage on the southeast side of Bligh Island, which had already been linked to the observatory. The survey was to be undertaken in an anticlockwise direction until they reached the open sea. The two launches duly departed on 18 August.

Meanwhile, in Friendly Cove, observations were taken in hand to obtain the rates of the chronometers and the longitude of the observatory. The first set of observations proved useless for rating the chronometers, but when Malaspina was sure of the rate of the astronomical clock, he recorded the results in his journal as follows:

| | *Chronometer 71* | *Chronometer 72* | *Pocket chronometer 105* |
|---|---|---|---|
| Longitude E of Mulgrave | 13°9′59″ | 12°55′54″ | |
| Therefore from Paris | 129°36′36″ | 129°50′41″ | 129°33′45″ |

The resulting mean longitude of 129°40′21″ differed by more than a degree not only from the results of the lunar distances observed in June and July but also from those that were observed on the following two days, 401 sets having been taken ashore in the best conditions, giving a longitude of 128°26′ west of Paris. Meanwhile Vernacci and Concha determined the latitude of the observatory by observing the meridian altitudes of different stars to the north and south of the zenith. The observations of the pendulum were undertaken by alternating groups of officers, until there was no doubt of the exact relationship of its oscillations to the mean time of the observatory.

On the 25th, Espinosa and Cevallos returned after an absence of eight days. They had established that as many as five channels or arms, usually no more than a third of a mile wide, ran inland in different directions, ending in small bays chosen by the Natives for their settlements or habitations. Muchalat and Tlupana Inlets, the two channels to the east and northeast, the seaward ends of which had been investigated by Cook, finished at the two villages of Tlupananú, one of the minor chiefs. The channel leading to the sea via Esperanza Inlet divided into three branches to the north. Tahsis Inlet, the first or easternmost, ended at Tahsis, the residence of Maquinna; Zeballos Inlet, the second branch, led toward the village of the subordinate chief, Natzapé; and Espinosa Inlet, the third branch, although no smaller in size, appeared to be deserted along both shores and around the bay at the end.[49] In general, no bottom could be reached in sixty fathoms in any of the channels, even near the shore. On their return to Friendly Cove from Esperanza Inlet via the open ocean, the two officers did not survey the large opening of Nuchatlitz Inlet. They did, however, fix the position of Inner Bajo Reef off Bajo Point. In 1778 Cook had probably seen the larger Bajo Reef, which is considerably farther out to sea. The fact that only a single reef was placed on the charts of the respective navigators has led to some confusion, but it seems clear that they identified different parts of the shoal. During their survey, Espinosa and Cevallos took daily observations for latitude, longitude, and variation, linking the positions together with numerous bearings, either taken on shore or confirmed by bases run by log-line. Finally, at Port Eliza, at the seaward end of Esperanza Inlet, they were unable to measure a base by chain. Instead, a base was measured not only by log-line but also by the speed of sound – determining by chronometer the time between the discharge and hearing the shot, repeated twice; the two methods agreed closely.[50]

In the details of Malaspina's astronomical observations given by Espinosa in 1805[51] and repeated by him four years later,[52] the latitude of the observatory is given as 49°35′15″ N and its longitude by the three chronometers as 13°05′20″ east of Port Mulgrave. By applying the accepted longitude of Port Mulgrave of 133°24′12″ west of Cádiz, the longitude of Friendly Cove was established as 120°18′52″ west of Cádiz. The observations

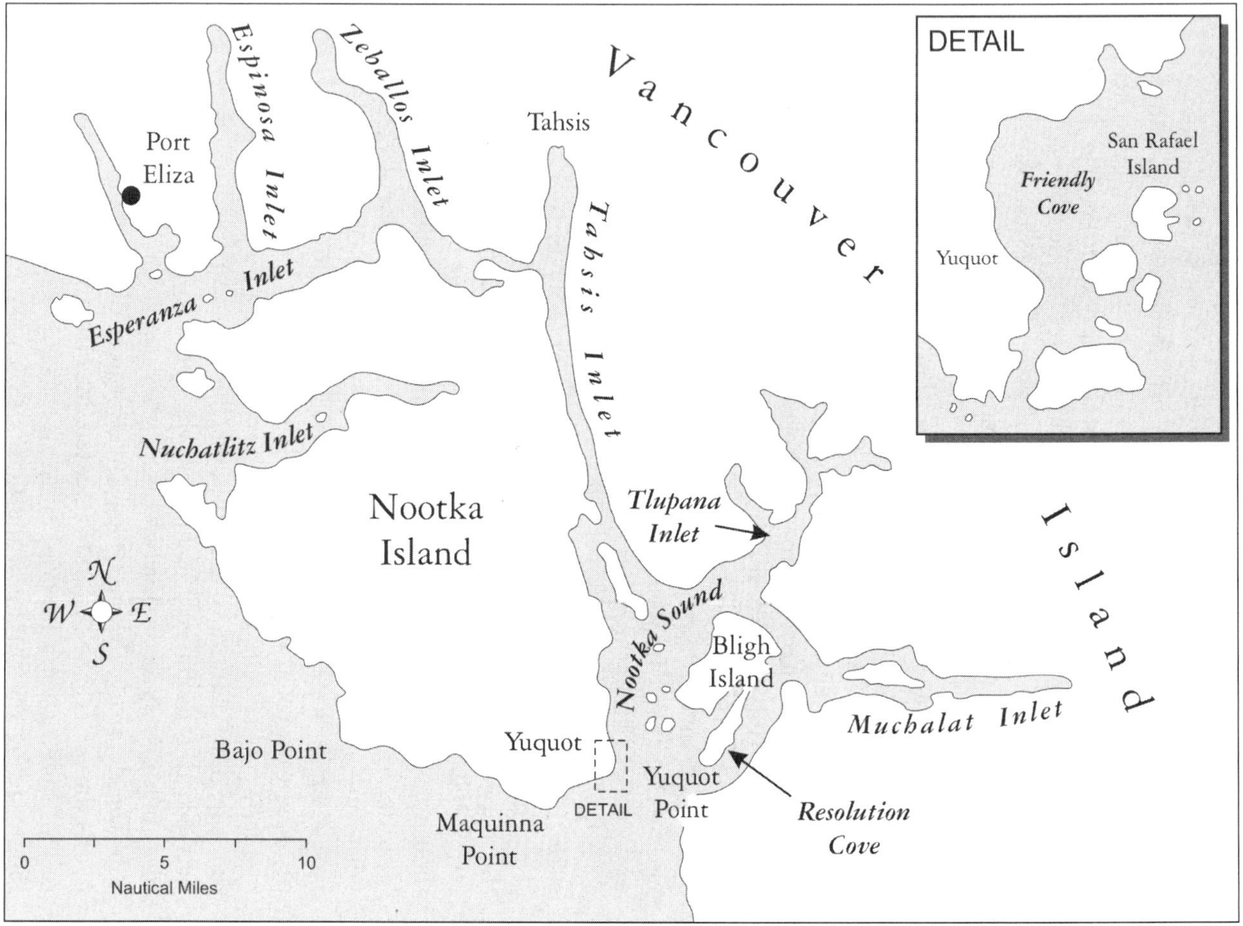


Sketch map of Nootka Sound today (Map: Chris Johnson)

for longitude by lunar distances were not taken into account, possibly because of their dubious nature. More accurate observations for longitude were obtained when Galiano visited Friendly Cove in 1792 in the *Sutil* and the *Mexicana*.[53]

Malaspina's survey of Friendly Cove was published in 1802 as "Plano de la cala de los amigos situada en la parte ocidental de la entrada de Nutka – año 1791,"[54] but for some unknown reason, the position of the observatory given on the plan is slightly different again, being 49°35′13″ N, 120°23′13″ west of Cádiz. Espinosa's and Cevallos's survey of the interior channels was not published, but when Vancouver visited Friendly Cove in 1792, a copy of this survey was apparently one of a number of Spanish surveys presented to him by Bodega y Quadra.[55] Nevertheless, when Vancouver's charts were finally engraved, Nootka Island was firmly joined to the main island,[56] in spite of the fact that Joseph Baker's manuscript chart covering this area incorporates Espinosa's and Cevallos's survey.[57] A possible explanation may lie in a copy of a Spanish manuscript chart of Manuel Quimper's 1790 expedition, forwarded by the Admiralty to Vancouver in the *Daedalus*.[58] This chart also depicts Nootka Island joined to the main island and may have been among the charts that Baker was working on after the voyage; it thus could well have been the source of this error.[59]

On his departure from Nootka on 28 August, Malaspina made for Acapulco, which he reached on 19 October. Here he found that Conde de Revillagigedo, the viceroy of New Spain, was preparing to send an expedition to complete the Spanish explorations of the Strait of Juan de Fuca under the veteran mariner Francisco Mourelle in the schooner *Mexicana*, which had just been built at San Blas. But on his arrival in Acapulco, Malaspina proposed to the viceroy that the expedition should be enlarged and commanded by Galiano in the schooner *Sutil*, also just completed in San Blas, with Cayetano Valdés in command of the *Mexicana* with Secundino Salamanca and Vernacci respectively as their seconds-in-command. The viceroy consented, and Mourelle brought the two schooners from San Blas to Acapulco so that they could be fitted out there under Malaspina's directions.

The schooners were well supplied with astronomical and surveying instruments, which Malaspina listed. They included the

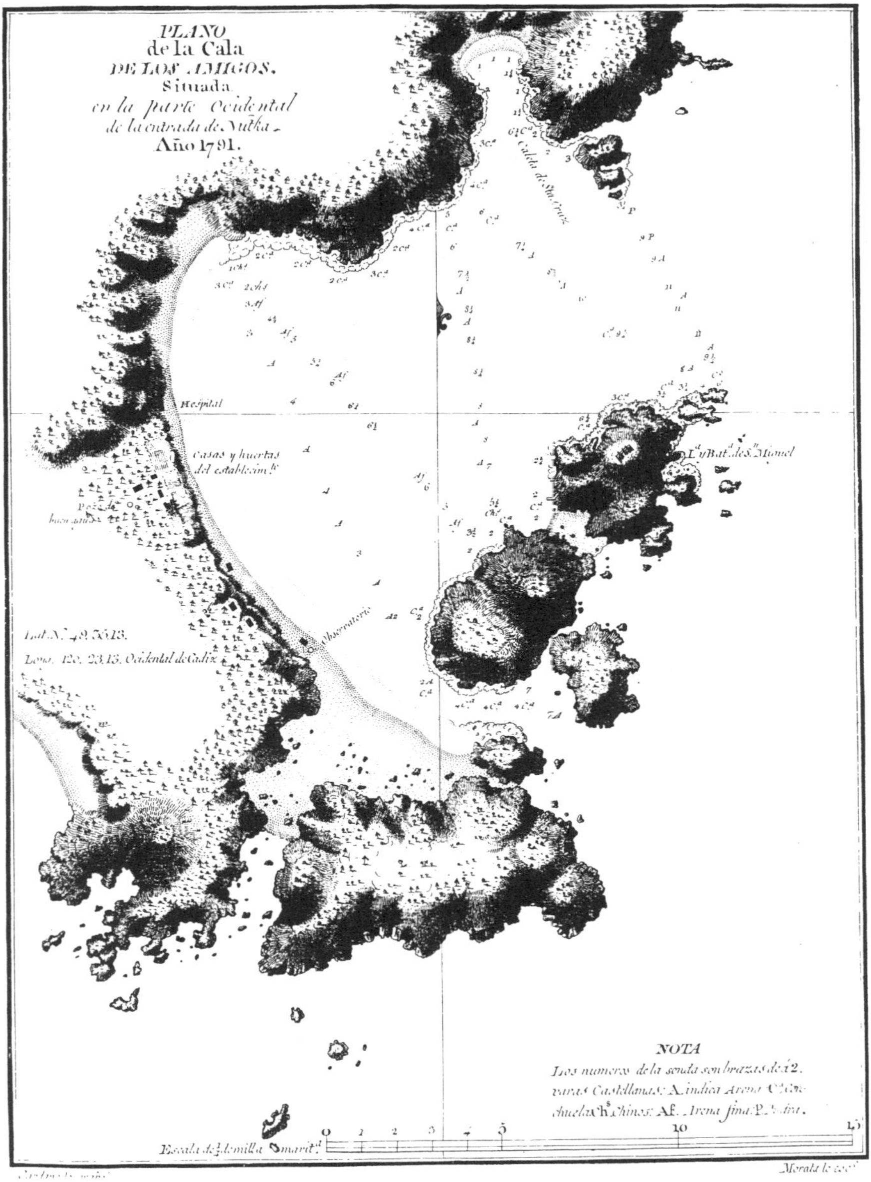


Plano de la Cala de los Amigos. (From *Relación del viage*)

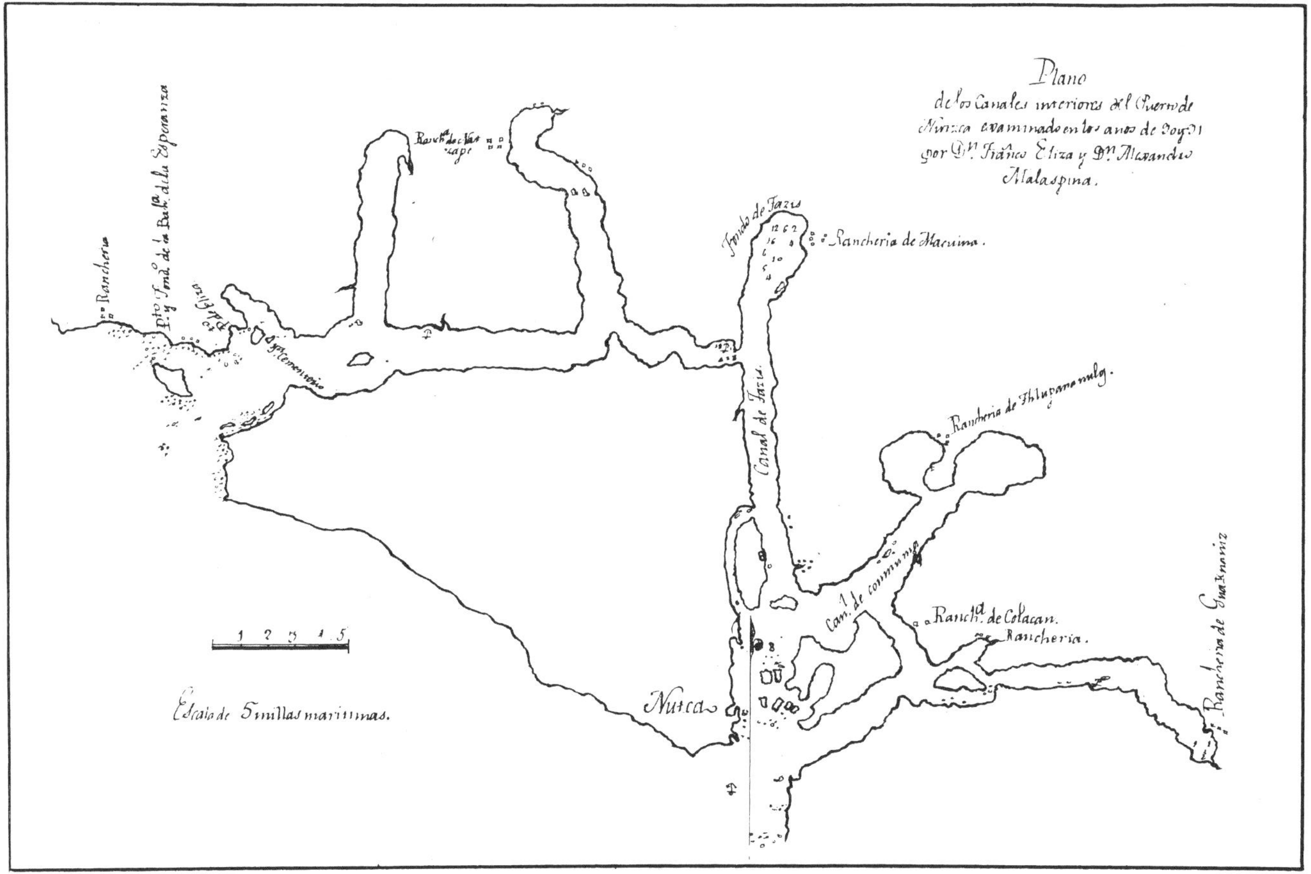


Plano de los Canales interiores. (From *The American West Coast and Alaska*; reprinted with permission of H.P. Kraus)

astronomical quadrant by Adams and the Ellicott astronomical clock, which Malaspina had found in Mexico City, as well as the large achromatic telescope and the pocket chronometer Arnold 344 that were already in Galiano's charge. To these were added chronometer 61, recently repaired in Mexico City, a second achromatic telescope, a barometer, some thermometers, a reflecting circle with a stand, a theodolite, an azimuth compass, and a mariner's compass. The schooners were also supplied with various charts, including not only the expedition's own surveys but also those of Captains Meares and Dixon and, in particular, the survey made in 1791 by Eliza in the Strait of Juan de Fuca.[60] The officers took with them their personal sextants, and it is known they also had three artificial horizons. They would certainly have carried surveyor's chains of one hundred English feet as well.

Galiano made his departure from Acapulco on 8 March 1792, setting a course directly for Nootka Sound. Though forced to keep well offshore to benefit from the northeast trade wind, he finally anchored in Friendly Cove on 13 May. On his arrival there, his first concern was to obtain an accurate value for longitude by observing the emersions of the satellites of Jupiter since, as mentioned earlier, the lunar distances obtained by Malaspina had proved unreliable. On 18 May a longitude of 120°30′15″ west of Cádiz was obtained by the emersion of Jupiter's first satellite, and on 28 May, of 120°49′15″ by the emersion of Jupiter's second satellite; given the difference of 19′00″, these results were not very encouraging. Galiano sailed for the Strait of Juan de Fuca on 5 June, and after brief stops at Núñez Gaona (Neah Bay), where the Spanish were setting up an establishment in case the one in Nootka Sound had to be abandoned, and Puerto de Córdoba (Esquimault Harbour), he anchored off Lopez Island, one of the San Juan Islands, in time to observe the emersion of Jupiter's second satellite. This gave the longitude of Nootka as 120°26′00″ west of Cádiz, the difference in longitude between Lopez Island and Nootka presumably having been measured by chronometer. Once again this value did not appear to be very satisfactory. However, from early in the voyage Malaspina and his officers had been aware that predicted values for lunar distances and the occultations of Jupiter's satellites were subject to errors; they realized that if the same phenomena had been observed in observatories in Europe, a much more accurate value for longitude could be computed at the end of the voyage. This happily proved to be the case with Galiano's observations, which had also been observed by Charles Messier in Paris. As a result, corrections of 9′30″, 29′45″, and 9′30″ had to be applied to Galiano's longitudes, giving a mean value for the longitude of Nootka of 120°19′00″ west of Cádiz[61] or 126°31′16″ west of Greenwich, compared with its current accepted longitude of 126°37′02″.

Very little is said about Galiano's surveying activities during the course of his voyage in the ensuing publication,[62] but what is said is significant. By the time he joined the *Sutil*, he was also convinced that obtaining longitude for surveying purposes by lunar distances had its drawbacks; he pointed out that errors he had experienced in lunar distances could

> be attributed almost entirely to an error in the tables with which the distances were computed. Thus, far from agreeing that longitude by lunar distances can be relied on to a quarter of a degree as had been advanced by some, one should only rely on them to the said three quarters, accepting that in most cases the mean of various series could be within one quarter of a degree. So we told Captain Vancouver, to whom our proposition was strange because of the ideas established in England by the best astronomers, who had predetermined, as an exact method of establishing longitude, the mean of many lunar distances.[63]

The voyage of the *Sutil* and the *Mexicana* and Galiano's meeting with Vancouver have been related in detail elsewhere and need not be repeated here, but in discussing the chart that was published on completion of his voyage,[64] Galiano pointed out that it

> has all the accuracy permitted by the narrow inlets with precipitous sides, where astronomy with observations must supplement geodesy. The artificial horizons had been in constant use, as much for latitudes as for longitudes by the chronometers, and after the longitude corresponding to the observation of the satellite of Jupiter at the anchorage at San Juan had been taken, all the longitudes on the coast had been corrected to the results of this observation, being preferable to all those made in this year and the previous one during the expedition commanded by Don Alejandro Malaspina.[65]

When Malaspina returned to Cádiz in September 1794, he was well received and promoted to *Brigadier de Real Armada*. At the end of November he travelled to Madrid, where he at once began preparing the account of his voyage. This publication was going to be on a grand scale, dwarfing in size and comprehensiveness the narratives of his predecessors in the Pacific, particularly Bougainville and Cook. There were to be no fewer that seven volumes, including Malaspina's own journal, the voyage of the *Sutil* and the *Mexicana,* the astronomical results of the voyage, the political state of Spain's overseas empire, observations on health at sea, and so on. Additionally, there was to be a volume of seventy drawings from the voyage and an atlas of seventy charts. However, Malaspina became involved in an attempt to replace Manuel Godoy as chief minister of Spain. As a result, he was arrested on 22 November 1795, tried, stripped of his rank, sentenced to life imprisonment, and incarcerated in Castillo de San Antón at La Coruña. After almost seven years in this fortress prison, he was released on the intervention of Napoleon and exiled to his native country, where he died in April 1808.

As a result of Malaspina's disgrace, the proposed publication was abandoned. However, in 1802, possibly as a counter to the appearance of Vancouver's *Voyage* in London four years earlier, an account of the voyage of the *Sutil* and the *Mexicana* was published as *Relación del viage hecho por las goletas Sutil y Mexicana en el año 1792 para reconocer el estrecho de Juan de Fuca*, without any reference to the Malaspina expedition as a whole or any mention of his name, "the commandant of the corvettes" or simply "the corvettes" being used instead. Many of the charts resulting from the expedition were also published, but again without Malaspina's name being given, just those of the *Descubierta* and the *Atrevida.* His charts of Yakutat Bay and Nootka Sound have already been mentioned; however, the *Atlas para el viage de las goletas Sutil y Mexicana al reconimiento del estrecho de Juan de Fuca en 1792* also included "Continuación de los reconocimientos hechos en la costa NO de América por los buques de S.M. en varias campañas desde 1774 á 1792," which incorporated Malaspina's running survey, combined with the surveys carried out by Martínez and Haro in 1788 in the *Princesa* and the *Filipo*, alias the *San Carlos*; by

Fidalgo in 1790 in the *San Carlos;* and by Jacinto Caamaño in 1792 in the *Aránzazu.*[66]

The embargo on mentioning Malaspina's name did not last very long; nor did a complete ban on publishing an account of his voyage. In 1809 Espinosa included a 224-page account of the voyage in his *Memorias*, which, though primarily concerned with Malaspina's astronomical and gravity observations, nevertheless also gave a comprehensive account of the voyage.[67] After a copy of Espinosa's *Memorias* reached St Petersburg, a 15-page summary of Malaspina's voyage was published in 1815 by Admiral Ivan (Adam) Fedorovich Kruzenshtern,[68] who was particularly interested in the exploration of the Pacific. So when a copy of Malaspina's journal was purchased by a Russian diplomat in Madrid a few years later, Kruzenshtern arranged for its publication in St Petersburg between 1824 to 1827 in a number of parts in the journal *Zapiski gosudarstvennogo admiralteiskogo departamenta.*[69] Thus the first full account of Malaspina's voyage appeared in Russia and not in Spain. It was not until 1885 that the first Spanish edition of his voyage was published in Madrid, under the editorship of the distinguished naval historian Pedro de Novo y Colson. It thus became widely available for the first time almost one hundred years after the expedition's return to Spain.

# 3

# *The Publication of British Admiralty Charts for British Columbia in the Nineteenth Century*

ANDREW S. COOK

Accurate topographical knowledge is a prerequisite for navigating with safety on the coast of northwestern North America.[1] Surveys and observations published as charts and sailing directions have always been the means of coordinating this coastal knowledge, particularly where communications use complex patterns of channels and passages. In British Columbia in the nineteenth century, topographical intelligence was gathered primarily by officers of Royal Navy vessels on specific survey work, and the results compiled and published by the Hydrographic Office of the Admiralty in London, until the transfer in 1910 of charting responsibilities to the Royal Canadian Navy and the new Canadian Hydrographic Service. The first dominion government survey instructions for the west coast were for Vancouver Harbour, compiled by W.J. Stewart in 1891, and CGS *Lillooet*, launched on the west coast in 1908, first operated on survey work in Dixon Entrance in 1909–10. But in the century following George Vancouver's exploration, the officers and ships whose work gave names to coast and sea features were primarily those of the Royal Navy.[2]

The history of hydrographic survey on the Northwest Coast is closely connected with naval and political history. A survey vessel was a naval presence, and a Royal Navy ship visit was a survey opportunity. Naval officers were under instructions to keep remark books, take views, and make plans; surveying officers were aware that they could be invoked by the civil administration or, in less accessible places, initiate the only effective authority.

After James Cook's landfall at Nootka Sound in 1778, Vancouver's surveys in 1791–94 were the earliest coherent body of topographical intelligence of what is now Vancouver Island and the adjacent coast.[3] Even before the Nootka Sound incident in 1790, merchants from India, and from Britain with licences from the South Sea Company and the East India Company, traded in sea-otter pelts from the Northwest Coast to Canton.[4] They knew the places they visited, and some delivered sketches to Alexander Dalrymple, hydrographer to the East India Company, for publication among his charts.[5] James Colnett, of the Nootka Sound Company, assembled his observations into a recognizable depiction of coastal features north of the Strait of Juan de Fuca, though it remained a rough manuscript in the Hydrographic Office.[6] In Dalrymple's 1807 worldwide survey of available charts,[7] the northernmost chart for the west coast of North America (apart from Arrowsmith's Pacific Ocean map) was Broughton's survey of the Columbia River, published by Aaron Arrowsmith as "River Oregan" in 1798.[8] Naval vessels visiting the coast, such as *Phoebe*, *Racoon*, and *Cherub* in 1813,[9] left no charts, and the next recorded surveys came from the Hudson's Bay Company at Fort Vancouver (now Vancouver, Washington) on the Columbia River. Three charts, of the Columbia River, Observatory Inlet, and the Gulf of Georgia, produced in 1825 by Henry Hanwell Jr in *William and Ann*, the first HBC vessel on the coast, no longer survive, but may have been used in Arrowsmith compilations.[10] Aemilius Simpson, based at Fort Vancouver, surveyed the mouths of the Columbia River and the Fraser River in 1827, the latter for the company post at Fort Langley, and other harbours northwards to the Nass River. His sketch of the Fraser River was eventually published in 1849 as Admiralty chart 1922, part of a group that also used Cook's and Vancouver's earlier work for Vancouver Island.[11]

The first naval officers in the nineteenth century with specific instructions to make observations on the Northwest Coast were Edward Belcher and Henry Kellett, in *Sulphur* and *Starling*, during Pacific cruises from Callao in 1836–42. The instructions to Belcher included the following: "Political circumstances have invested the Columbia river with so much importance, that it will be well to devote some time to its bar and channels of approach, as well as to its inner anchorages and shores."[12]

Belcher had made a plan of Friendly Cove, in Nootka Sound, in 1837, after earlier surveying Port Etches, in Prince William Sound. His arrival in *Sulphur* at the mouth of the Columbia River in 1839 was less fortunate: Kellett, waiting in *Starling* to guide *Sulphur* in, grounded on the bar in the tide and had to limp upriver to Fort Vancouver for repairs. Belcher attributed the mishap to poor directions: "Fortunately the weather admitted of our entering, otherwise the very imperfect sailing directions might have led us into danger. The shoals in the entrance of this river have most materially changed their features within the last two years."[13] Before the survey that Belcher then instituted, there were for the Columbia River only Arrowsmith's 1798 chart and a small plan by Anthony Robson, "Commander of a Merchant Ship in 1816," issued from the Admiralty in 1821.[14]

Admiralty charts in the early nineteenth century depended on the initiative of officers such as Belcher carrying out surveys at Nootka Sound or the Columbia River, or on the far-sightedness of mariners such as Robson contributing their manuscript surveys. The Admiralty Hydrographic Office had developed from the position of hydrographer, created in 1795 to provide official status for Alexander Dalrymple's navigational and hydrographic advice to government.[15] The hydrographer was "to be intrusted with the custody and care of such plans and charts as now are, or may hereafter be, deposited in this office belonging to the public, and to be charged with the duty of selecting and compiling all such information as may appear to be requisite for the purposes of improving the navigation."[16] It was left to Dalrymple's successor, Thomas Hurd, to initiate the practice of issuing folios or atlases of charts from official and commercial sources together. Hurd established a survey service in 1817,[17] but for distant coasts the hydrographer could only hope to influence the secretary to the Board of Admiralty to include survey work in instructions to expedition officers.

Early sales catalogues of Admiralty charts (the public sale of charts commenced in 1821)[18] show the sporadic coverage of the world's coasts resulting from the necessarily passive involvement of the Hydrographic Office except where surveys had been specifically commissioned. Matthew Flinders's work in Australia, W.F.W. Owen's on the east coast of Africa, and W.H. Smyth's in the Mediterranean, each resulting in coherent chart series, contrast with the lack of charts for the west coast of North America. Francis Beaufort, hydrographer from 1829 to 1855, greatly extended the system of publishing and correcting series of coastal charts: for example, Henry Wolsey Bayfield's Gulf of St Lawrence charts and the revised series by J.F.W. DesBarres and Anthony Lockwood for Nova Scotia. No such progress towards reliable chart coverage could happen for the west coast of North America while visits of naval vessels were infrequent and their observations uncoordinated. By 1839 the Hudson's Bay Company dominated the fur trade against Russian and American rivals, partly as a result of the Russian-British convention of 1825, without requiring a guaranteeing naval presence. Belcher's concern had resulted in a single publishable survey.[19] The sloop *Modeste*, commanded by Thomas Baillie, visiting the Columbia River, Fort Victoria (the HBC's new establishment on Vancouver Island), and Port Simpson on the northern coast in 1844, demonstrated a concern to protect British interests in the Oregon territory. The *Modeste* too grounded on the Columbia River bar, but no surveys resulted.[20]

With the Oregon dispute settled in 1846, Fort Victoria (now Victoria) eventually replaced Fort Vancouver as the company's western headquarters. The agreement carried the boundary along the 49th parallel to the middle of the Strait of Georgia and then by the centre of the channel leading to the Pacific. The resulting need to survey the islands and inlets between Vancouver Island and Puget Sound brought the barque *Herald* and the brig *Pandora* to the Strait of Juan de Fuca in 1846. Kellett's surveys of Victoria, Becher and Pedder Bays, Sooke Inlet, and Port San Juan in 1846–47 went to the Admiralty with Lieutenant James Wood's survey of Esquimalt Harbour. All were published in 1848 as the first Admiralty charts of Vancouver Island,[21] and Kellett's recommendation helped to establish Esquimalt as the naval headquarters, close to Victoria, the seat of colonial government. Kellett's work in the Strait of Juan de Fuca in 1847 was compiled, with details from the United States

Exploring Expedition maps of 1841, into a chart of the strait in early 1849, to complement the harbour plans.[22]

Beaufort in February 1849 published a general chart of Vancouver Island and the Gulf of Georgia, derived primarily from Kellett's surveys and completed through Johnstone Strait to Queen Charlotte Sound and along the west coast of the island from Vancouver's surveys (dated to 1793) and those of the Spanish explorers Galiano and Valdés (dated to 1792).[23] At the same time he published Kellett's 1839 Friendly Cove plan as an inset to a chart of Nootka Sound, otherwise entirely in hairline from Cook's and Vancouver's surveys and from a Spanish manuscript, all long since registered in the Hydrographic Office archive.[24] Next, Simpson's 1827 Fraser River sketch was engraved from the archive,[25] and Beaufort also published R.M. Inskip's 1846 sketch of Puget Sound,[26] the nearest sea approach to Fort Vancouver for vessels too large to cross the Columbia River bar. The discovery of coal at Beaver Harbour in Johnstone Strait and at Nanaimo resulted in harbour plans – Beaver Harbour in July 1851 and November 1852 and Nanaimo first in 1856[27] – as the number of auxiliary steam vessels increased.

To exclude American territorial claims from the Queen Charlotte Islands when gold was discovered there in the early 1850s, the frigate *Thetis*, commanded by Captain Augustus Kuper, was sent from Callao in 1852, followed by the paddle-wheel sloop *Virago*, under Commander James Prevost, to maintain British sovereignty. Moore, of the *Thetis*, surveyed Mitchell and Douglas Harbours on the west coast and the "Gold Harbour" (or Port Kuper), which was the focus of the prospectors' interest.[28] Beaufort engraved the plan in 1853 and, for want of another survey to publish with it, included on the plate a hairline map of the Queen Charlotte Islands derived solely from Vancouver's chart, drawn sixty years earlier.[29] Three years later Beaufort had received sufficient sketch surveys of harbours and inlets, particularly from G.H. Inskip, master of the steam vessel *Virago*,[30] to cancel Vancouver's outline map and to reissue the plate as "Plans of Ports in the Queen Charlotte Islands" in November 1856.[31] Inskip's 1853 *Remarks for Sailing Directions*, published in 1856,[32] gives insights into the local fur trade: "Bears (which are very numerous), martens, sea and land otters, are caught for their furs, which are taken to the H.B.Co's.

establishment, at fort Simpson (port Machlochlin), or to fort Rupert (Beaver harbour) in Vancouver island."

Inskip and his fellow officers Gordon and Knox surveyed Port Simpson in 1853, for publication early in 1856.[33] Later in that year appeared a new general chart of the Queen Charlotte Islands, including the adjacent coast from Vancouver Island to Cordova Bay, using a Russian chart and corrections by Inskip to modify Vancouver's survey.[34] Deliberately intended as an "Index Chart" to a group of large-scale plans of harbours, it demonstrates the shortcomings of the survey data supplied to the hydrographer. But by 1856, throughout the Vancouver Island colony area, the standard pattern of Admiralty chart provision was emerging: small-scale coast or archipelago charts, often compiled "in office" from various sources, serving as keys to an increasing number of single-source large-scale plans.

Neither Beaufort nor, after January 1855, John Washington could dictate or predict what materials they would receive to work with. With the exception of Kellett's coasting survey in the Strait of Juan de Fuca, they had to use older or foreign materials as a coast framework for the new harbour surveys. The resulting charts, though conforming to the pattern of small-scale coastal compilations, gave only fragmentary information for mariners.

This situation changed after 1856 with the dispute between Britain and the United States over the San Juan Islands. The Oregon Treaty did not take into account the San Juan Islands, which divide the waterway connecting the Straits of Georgia and Juan de Fuca into three routes, rather than one channel through which a boundary could run. The Americans claimed the islands on the grounds that the intended channel was the westerly Haro Strait, between the San Juan Islands and Victoria; Britain maintained, from shipping usage, that the intended channel down which the boundary should be drawn was Rosario Strait to the east between the islands and the mainland, thereby attaching the islands to the Vancouver Island colony. Prevost, commanding *Satellite*, was appointed to the Anglo-American boundary commission in 1856, and Captain George Richards in the steamer *Plumper* arrived from surveys in New Zealand and in the Arctic as chief surveyor and astronomer. Richards's first survey was of Semiahmoo Bay,[35] to define where the 49th parallel cut the shoreline. He was continuously employed in surveys in the Vancouver Island colony from 1857 to 1863,[36] and from his work and the subsequent surveys of his assistant, Daniel Pender, developed the corpus of Admiralty charts of Vancouver Island and British Columbia.

Richards's first task in 1858 and 1859 was to survey the Haro and Rosario Straits, with the disputed San Juan Islands; the fair copies (using Kellett's work on the south shore of the Strait of Juan de Fuca) were sent to London, and the resulting chart 2689 published on 28 July 1859. To quicken publication, the detail of the Fraser River, still under survey by Richards in 1859, was first left blank. Simpson's 1827 sketch remained current until Richards's new survey was published in November 1860 as chart 1922 for the Fraser River and Burrard Inlet. The connecting river detail was immediately reduced for subsequent issues of chart 2689. During his survey, Richards was called in January 1859 to the Fraser River to help quell a disturbance near Fort Yale and sent his lieutenant, Richard Mayne, upriver with the sternwheeler *Enterprise*.[37] Mayne's survey of the river from Langley to Yale was lithographed in London in 1859 and issued in that year only as chart 2666: the river above Macmillan Island was not navigable by vessels with deeper draught.

With a plan of Nanaimo Harbour and Departure Bay (chart 2512, September 1860), contiguous larger-scale plans of Esquimalt Harbour and Victoria Harbour replacing the 1848 plans and using land topography from J.D. Pemberton's new surveys (charts 1897a and 1897b, December 1861),[38] and a coastal chart of Haro Strait and Middle Channel (chart 2840: November 1861), Richards completed the surveys of the southern part of the Strait of Georgia. To accompany the charts, he had prepared Part 1, so styled, of the *Vancouver Island Pilot*,[39] which was issued on 15 August 1861, following the new *Instructions ... [for] Sailing Directions*.[40]

The second phase of Richards's work, first in *Plumper* and then from 1861 in the larger paddlewheel sloop *Hecate*, continued northwards from Nanaimo and Burrard Inlet, following Vancouver's earlier passage. Two small-scale charts 579 and 580 for the Strait of Georgia and 581 for Johnstone and Broughton Straits from Knox Bay to Goletas Channel, published as a group in late 1862 and early 1863 and covering the complex channels from the Fraser River through Discovery Passage to Queen Charlotte Strait, were the result of Richards's 1860 season, though he left unsurveyed the areas north and east of the main channel from Quadra Island to Knight Inlet and beyond. The charts were accompanied by two plates of harbour plans, from Comox and Nanoose to Alert Bay and Beaver Harbour.[41]

Richards's work in 1861–63 on the west coast of Vancouver Island continued the series of coastal charts from Goletas Channel to Barkley Sound, to connect with the earlier survey in the Strait of Juan de Fuca.[42] From his surveys, the Hydrographic Office published larger-scale plans of Quatsino Sound, Klaskino and Klaskish Inlets and Anchorages, Nootka Sound, Nasparti and Ou-ou-kinsh Inlets, Kyuquot Sound, Esperanza and Nuchatlitz Inlets, and Barkley Sound itself.[43] The completed survey of the coasts of Vancouver Island merited the republication of *The Vancouver Island Pilot* in October 1864, now "for the whole of the coasts of Vancouver Island, including Juan de Fuca Strait, and the shores of British Columbia as far north as the parallel of 51° N., with the exception of some of the deep inlets."[44] Richards's new chart 1917 of "Vancouver Island and adjacent shores of British Columbia" in September 1865 added his coastal surveys to Kellett's work and to that of the United States Coast Survey for the Washington territory shore.

The delay between survey and chart publication, routinely as much as two to three years, encouraged Richards to experiment unofficially with printing fair surveys at the Royal Engineers' lithographic press at New Westminster. Unlike many hydrographic surveyors on distant coasts, he was fortunate in having colonial and military powers available ashore. He had 120 copies printed at New Westminster of his 1862 resurvey of Nanaimo Harbour, for steamers bunkering and other coaling vessels, but none appear to have survived.[45] Three similar Royal Engineers charts have been found, all of Barkley Sound and Alberni Canal, an area of developing commercial shipping.[46]

Recalled to Britain in 1863, Richards learnt of John Washington's death and his own promotion to hydrographer en route from the Pacific. He subsequently published much of his own Vancouver Island work and plans of Griffin Bay and Roche Harbour, respectively the American and British garrison settlements on San Juan Island, and of Duncan Bay and Metlah-Catlah Bay, the missionary settlement established by William Duncan after a smallpox epidemic at Port Simpson in 1862.[47] Other charts from Richards's own surveys were chart 555, of Goletas Channel and the passages to Queen Charlotte Sound, and chart 538, of Seymour Narrows in Discovery Passage, with the notorious submarine pinnacle Ripple Rock. But the charts of Vancouver Island and British Columbia published between 1864 and 1874 were chiefly those of Daniel Pender, Richards's surveying assistant until 1863. Pender, subsequently promoted staff commander, continued the surveys to 1870 in *Beaver*, a paddlewheel sloop hired from the Hudson's Bay Company,[48] chiefly on the mainland shore opposite Vancouver Island, from Toba and Bute Inlets northwards. He completed Richards's Texada Island to Johnstone Strait chart 580, and Richards republished the 1862 chart of Johnstone Strait in 1867 with the channels north of the strait completed. From patrolling Bute Inlet after a Chilcotin (Tsilhqot'in) massacre in 1864[49] came Pender's survey of Toba, Bute, and Loughborough Inlets, first as an extension panel to the head of chart 580. This made an unwieldy chart, and the northern part was transferred in 1900 to the new chart 2870.

From Pender's surveys came, in 1867, the two-sheet chart 1923a and 1923b of the open coast from Cape Caution to Port Simpson, with part of the Queen Charlotte Islands. His work in these northern waters, often extended in hairline from Vancouver's charts, with particular charts of Lama Passage and Seaforth Channel, Brown and Edye Passages, Ogden Channel, Port Simpson, and Nass Bay,[50] completed the Hydrographic Office coverage of the coast. His 1866–70 survey of the Queen Charlotte Islands, which also generated chart 48 of Skidegate Inlet in 1872, served, with G.M. Dawson's observations for the Geological Survey of Canada, as the basis of the new small-scale chart 2430 of the islands and Hecate Strait in 1882. The sailing directions that resulted from Pender's work (and incorporating Inskip's Queen Charlotte Islands *Remarks*) were published in June 1883 as an extensive *Supplement* to the 1864 *Vancouver Island Pilot*.[51] The two books were combined in March 1888 in the first edition of *The British Columbia Pilot*.[52]

Richards's achievement in the Hydrographic Office was in developing systems for processing survey data quickly and efficiently, both in using it for new charts, and in applying it to correct existing charts.[53] *Notices to Mariners* had long existed for publishing supplementary navigational information, but the chart plates were only infrequently corrected: when a new survey was received of a harbour, a new plate was usually engraved. Only where series of office-compiled charts were in general use – for example, in the English Channel – was piecemeal correction the norm. Newly observed dangers could be announced by notice, but to enter them on existing charts, where those charts had not been used to locate them, was uncertain. Even so, charts from the late 1840s were increasingly reissued with "Additions" or "Corrections" supplementing the publication imprint. The spread of coastal-chart cover encouraged the use of charts as the basis for reporting corrections as well as for making revisions or extensions. From 1866, the month and year of small corrections in *Notices to Mariners* were printed at the lower left corner of charts, supplementing the major revisions already recorded as a "Corrections" line at the right of the imprint.[54] Richards's own charts of Vancouver Island were among the first to be corrected in this way with Pender's extension surveys, revisions, or simple corrections. In 1867 Richards systematized the publication of *Hydrographic Notices* as supplements to the increasing range of published *Sailing Directions* or Pilots.[55] Remarks, such as those of Inskip for the Queen Charlotte Islands, had been issued unsystematically since the 1850s. Richards published *Hydrographic Notices* in annual series from 1867, numbering them also as cumulative series supplementing each book of sailing directions or each wider area still without formal *Pilots* (the Pacific Ocean before 1885 or Newfoundland and Labrador before 1878). The first *Hydrographic Notice* for Vancouver Island reported in September 1866 the discovery of Ripple Rock,[56] shown on new chart 538 for Seymour Narrows in the same year. The notice was reissued in May 1867 to supplement the *Vancouver Island Pilot* of 1864.[57]

Pender's surveys concluded in 1870 in Fitzhugh and Milbank Sounds on the open coastline north of Queen Charlotte Sound. Richards had pressed for them to be continued after the United States purchase of Alaska in 1867, despite colonial government reluctance to contribute to survey work, in order to establish the northern maritime boundary of British Columbia while the San Juan Islands dispute remained unresolved.[58] Responsibility for navigation and shipping matters had been transferred from Britain to the new dominion in the British North America Act of 1867, and the accession of British Columbia to Confederation on 20 July 1871 ended Britain's formal responsibility for west-coast surveys.[59] Pender returned to London, first under Richards and then as assistant hydrographer after Richards retired in 1874. For the German arbitration of the San Juan Islands dispute in 1871, Richards prepared the map of the Straits of Georgia and Juan de Fuca as a lithographic transfer from parts of charts 2840 extended westwards into the Strait of Juan de Fuca.[60] British interest in charting in British Columbia diminished, though minor investigations were carried out by ships from Esquimalt, and a new chart 1911 of the Strait of Juan de Fuca was published from "Admiralty Surveys to 1883." A new survey, by Lieutenant B.M. Chambers in *Nymphe* in 1895, of Menzies Bay, close to Seymour Narrows, appeared on the 1896 reissue of chart 538 alongside Richards's 1860 survey of the narrows. Charts were otherwise updated in London by means of Richards's "small corrections," and the archive of his and Pender's fair tracings in the Hydrographic Office[61] was used to construct new charts. Chart 1922 became inadequate for the industrial development spreading from Port Moody towards the new Vancouver, and an interim larger-scale chart 922 for Burrard Inlet had to be produced in 1886 from Richards's 1859 survey.

The commercial expansion of Vancouver and particularly the grounding of the Canadian Pacific liner *Parthia* in June 1890 provoked the first Canadian government survey on the west coast. The Canadian Department of Marine and Fisheries had begun surveying work only in 1883, and in November 1890 it detached William J. Stewart from the Georgian Bay Survey, under Staff Commander J.G. Boulton, to survey Burrard Inlet the following year.[62] With no chart publication yet in Canada, the results were published as the new Admiralty chart 922 of Burrard Inlet in March 1893. Changing commercial shipping needs prompted the publication in 1899 of Richards's forty-year-old survey from Active Pass to Gabriola Pass and the Gulf Islands, as steamship routes from Vancouver connected with the new railways of the Saanich Peninsula to serve Victoria.[63] Active Pass, later republished as a large-scale plan, had been found a

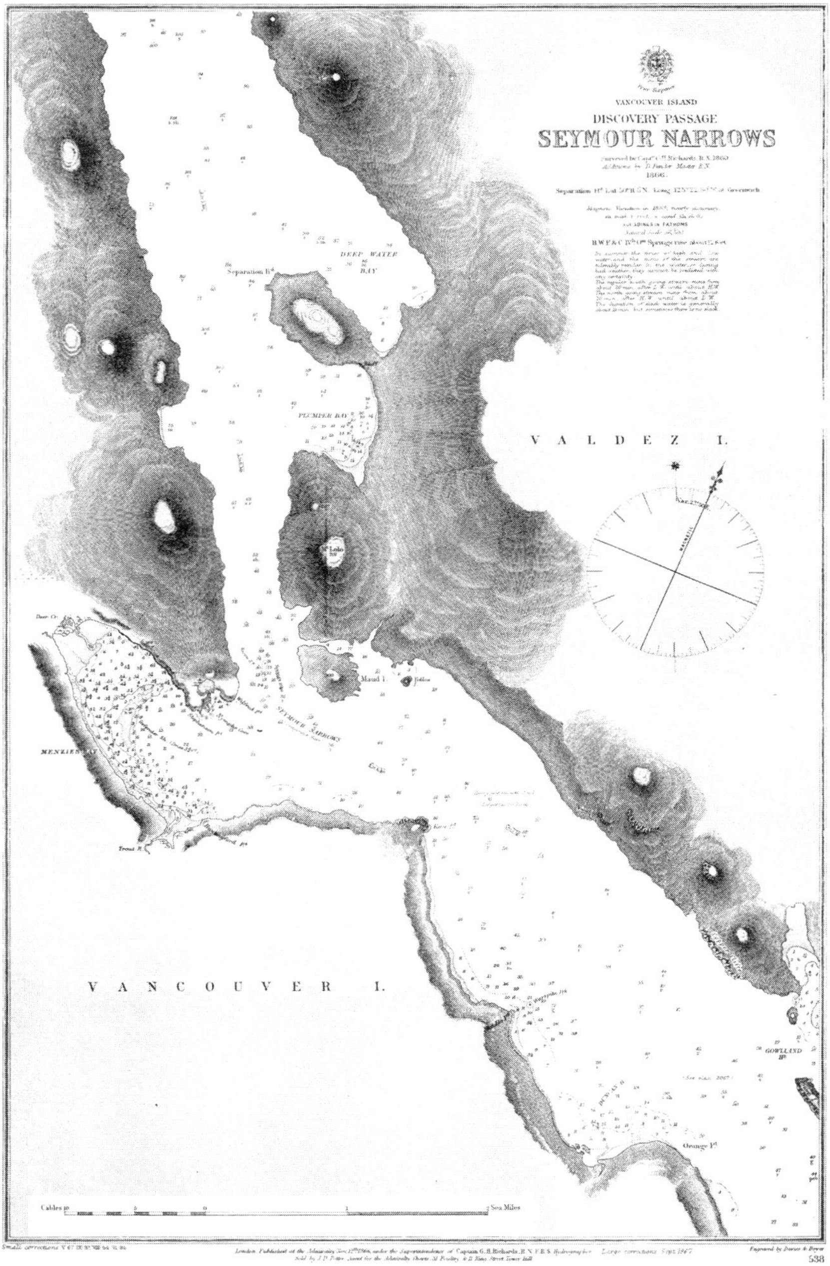


Admiralty Chart 538 of Seymour Narrows, published 12 November 1866 and corrected to June 1895 (see page 59)

reliable, though sinuous, deep-water passage through the protective outer chain of islands facing the Strait of Georgia. A voluntary contributor of plans and observations to the 1890s charts was John T. Walbran, from 1891 in command of the CGS *Quadra* servicing lighthouses and beacons and investigating reports of dangers.[64]

Increased commercial shipping in the Strait of Georgia in the late 1890s, coupled with advances in steam propulsion, and the scheduling demands of competing shipping companies alerted the Department of Marine and Fisheries in Ottawa to "the necessity of a more accurate survey of our Pacific Coast than has hitherto existed [because of] the many strandings that have occurred on uncharted dangers and by the frequent discovery of new dangers."[65] The Admiralty appointed Captain M.H. Smyth to the Pacific station at Esquimalt in 1898, commissioning the survey vessel *Egeria* to retriangulate and survey the Strait of Georgia for trans-Pacific telephone cables southwestward across Vancouver Island.[66] Charting standards had improved since Richards's 1858–63 surveys, and the two basic small-scale charts 579 and 580 of the Strait of Georgia latterly bore the note that their longitude scales required a correction factor for use together. *Egeria* remained on station until 1910, successively under Smyth, C.H. Simpson, J.F. Parry (twice), F.C. Learmonth, and J.D. Nares, producing surveys that resulted in nearly thirty new charts, besides inset plans. The first local requirement in 1898 was for surveys of Baynes Sound and the approaches to Comox; Baynes Sound was the port for coal from adjacent Cumberland and was used for naval gun and musketry practice, with the naval firing range at Comox, and two charts were consequently published in 1900.[67] Smyth also surveyed First and Second Narrows to supplement Stewart's 1893 chart of Burrard Inlet. Simpson resurveyed the inner route from Discovery Passage to Queen Charlotte Sound in 1900–02, for three charts of Johnstone Strait and plans from Alert Bay to Gowlland Harbour.[68] Both Simpson in 1903 and Parry in 1904 spent part of the surveying season at Nanaimo or in the northern Gulf Islands, and Parry's 1905 work between Moresby Passage and Gabriola Pass resulted in the two charts 3618 and 3619 appearing in mid-1907. Learmonth took *Egeria* to Port Simpson and the Queen Charlotte Islands in 1906 and 1907, and apart from resurveys in Moresby Passage and Chatham Sound in 1908, much of her service thereafter was in the waters of Dixon Entrance.

Before the survey vessel CGS *Lillooet* was launched in 1908, P.C. Musgrave (formerly a lieutenant in the Royal Navy) and G.B. Dodge, with H.D. Parizeau and L.R. Davies, surveyed Prince Rupert Harbour in 1906–07 for a railway port. Musgrave took *Lillooet* to contribute to survey work in Dixon Entrance in 1909 and 1910. The survey of Prince Rupert was published first as a Canadian Hydrographic Survey chart in October 1909 and as chart 3282 in the Admiralty series in September 1910.

*Egeria* was decommissioned in 1910 on the transfer of Royal Navy responsibility for protection and survey on the coast to the new Royal Canadian Navy. New surveys, by Canadian surveyors, went to the Hydrographic Survey of Canada in Ottawa to develop a new body of Canadian charts. Many Admiralty charts were transferred photographically into Canadian "frames" and reissued with Canadian chart numbers; information from Admiralty charts was incorporated, with new survey data, in new Canadian charts. New Admiralty charts were produced, but now from Canadian government surveys, often with different coverage and scales. There were still 120 British charts current for British Columbia in 1927. That number reduced to 40 by 1947, after a particularly heavy cull in 1942, when Admiralty rescheming of chart coverage of the west coast of Vancouver Island withdrew the inlet charts from the British portfolio and reduced the coastal to a sequence of 3. The government requirement for mariners to use Canadian charts in Canadian waters left, by the 1990s, fewer than 20 British charts to give international access to ports in British Columbia. These twentieth-century events are part of another history, that of the Canadian Hydrographic Service chart. The feature of the nineteenth century in British Columbia was the development in sixty years, from a zero base, of comprehensive charts, plans, and sailing directions for a developing navigation, achieved in two short periods: from 1858 to 1870 through the initiative of two men, George Richards and Daniel Pender, with the officers of *Plumper, Hecate,* and *Beaver*, and from 1898 to 1910 by the revising work of J.F. Parry and successive commanders and officers of *Egeria*.

Table 3.1

BRITISH ADMIRALTY CHARTS OF BRITISH COLUMBIA TO 1911

Admiralty charts published from 1848 to 1911 between Strait of Juan de Fuca and Dixon Entrance. "Surveyor and date" information is taken from the charts: square brackets indicate later revisions, and round brackets the incorporation of earlier surveys. "Current" signifies the years for which charts are listed in Admiralty Chart Catalogues.

| *Published* | *Chart* | *Title* | *Surveyor and date* | *Current* |
|---|---|---|---|---|
| 1848 Sep 8 | 1901 | Esquimalt Harbour | Wood *Pandora* 1847 | 1848–1861 |
| 1848 Sep 15 | 1897 | Victoria Harbour | Kellett *Herald* 1847 | 1848–1861 |
| 1848 Dec 1 | 1906 | Becher and Pedder Bays | Kellett *Herald* 1846 | 1848–1878 |
| 1848 Dec 8 | 1907 | Sooke Inlet | Kellett *Herald* 1847 | 1848–1866 |
| 1848 Dec 26 | 1910 | Port San Juan<br>[*insets*] Duncan Rock; Neeah Bay | Kellett *Herald* 1847 | 1848–1882 |
| 1849 Jan 18 | 1911 | Strait of Juan de Fuca | Kellett *Herald* 1847; U.S. Ex. Ex. 1841 | 1849–1883 |
| 1849 Feb 24 | 1916 | Nootka Sound<br>[*inset*] Friendly Cove | Cook; Vancouver; Spanish MS<br>Belcher *Sulphur* 1839 | 1849–1865 |
| 1849 Feb 28 | 1917 | Vancouver Island and Gulf of Georgia | Vancouver 1793; Galiano, Valdes 1792;<br>Kellett *Herald* 1847 [Richards 1859] | 1849–1865 |
| 1849 Apr 20 | 1922 | Fraser River | Simpson *Cadboro* 1827 | 1849–1860 |
| 1849 Oct 11 | 1947 | Puget Sound | R.M. Inskip 1846 | 1849–1883 |
| 1851 Jul 11 | 2067 | Port Shucartie; Western end of Beaver Harbour | Dillon *Daedalus* 1850 | 1851–1863 |
| 1852 Nov 10 | 2153 | Beaver Harbour | Mansell 1851 | 1852–1862 |
| 1853 Mar 11 | 2168 | Queen Charlotte Islands and adjacent coast<br>[*inset*] Port Kuper including Mitchell and Douglas Harbours | Vancouver<br>Moore *Thetis* 1852 | 1853–1855 |
| 1856 Feb 28 | 2426 | Port Simpson | E.H. Inskip, Gordon, Knox/Prevost<br>*Virago* 1853 | 1856–1871 |

Table 3.1 (continued)

| *Published* | *Chart* | *Title* | *Surveyor and date* | *Current* |
|---|---|---|---|---|
| 1856 Jun 25 | 2430 | Vancouver Id. to Cordova Bay | Russian chart 1849, Inskip 1855 [*later editions* Vancouver 1792, Russian chart 1849, Inskip 1854] | 1856–1880 |
| 1856 Nov 8 | 2168 | Plans of Ports &c. in Queen Charlotte Islands: | | 1856–1952 |
| | | Houston Stewart Channel and Rose Harbour; | E.H. Inskip, Gordon, Knox/Prevost *Virago* 1853 | |
| | | Virago Sound | E.H. Inskip 1853 | |
| | | Parry Passage [*deleted 1881*] | | |
| | | Entrance to Masset Harbour | Knox 1853 | |
| | | Entrance to Cumshewas Harbour | Sinclair H.B.C. 1852 | |
| | | Skidegate Bay [*deleted 1881*] | Sinclair H.B.C. 1852 | |
| | | [*inset*] Port Kuper including Mitchell and Douglas Harbours | Moore *Thetis* 1852 | |
| | | [*later insets*] Selwyn Inlet | Dawson 1878 | |
| | | Skincuttle Inlet | Dawson 1878 | |
| 1856 Nov 10 | 2512 | Nanaimo Harbour | H.B.C. officers; Inskip 1853 | 1856–1860 |
| | | [*inset*] Knox Bay | Knox 1853 | |
| 1858 Oct 11 | 2627 | Semiahmoo Bay and Drayton Harbour | Richards "1856" [i.e. 1857?] | 1858–1864 |
| 1859 May 16 | 2666 | Upper Part of Fraser River from Langley to Yale | Mayne/Richards *Plumper*, Begbie, B.C. Judge 1859 | 1859 only |
| 1859 Jul 28 | 2689 | Haro and Rosario Straits | Richards *Plumper* 1858–59 (Kellett 1847) | 1859–1882 |
| 1860 Sep 1 | 2512 | Nanaimo Harbour and Departure Bay | Richards *Plumper* 1859 [Smyth *Egeria* 1899] | 1860–1905 |
| 1860 Nov 30 | 1922 | Fraser River and Burrard Inlet | Richards *Plumper* 1859–60 [Stewart 1891] | 1860–1952 |
| 1861 Nov 21 | 2840 | Haro Strait and Middle Channel | Richards *Plumper* 1858–60 | 1861–1911 |
| 1861 Dec 13 | 1897a | Esquimalt Harbour | Richards *Plumper* 1858; Pemberton | 1861–1923 |

Table 3.1 (continued)

| *Published* | *Chart* | *Title* | *Surveyor and date* | *Current* |
|---|---|---|---|---|
| 1861 Dec 13 | 1897b | Victoria Harbour | Richards *Plumper* 1858; Pemberton | 1861–1923 |
| 1862 Sep 30 | 580 | N.E. Pt. of Texada I. to Johnstone Strait (Strait of Georgia sheet 2) | Richards 1860 | 1862–1925 |
| | | [*extended 1867–1900*] including Toba, Bute & Loughborough Inlets | Pender 1864 | |
| | | [*insets*] Waddington Harbour; Beaver Creek; Gorge Harbour; Drew Harbour; Squirrel Cove; Cameleon Harbour [*all deleted 1900*] | Pender 1863 | |
| | | [*later inset*] Tribune Bay | Smyth *Egeria* 1898 | |
| 1862 Dec 31 | 581 | Johnstone & Broughton Straits, Knox Bay to Goletas Channel | Richards 1860 | 1862–1867 |
| 1863 Feb 16 | 579 | Fraser R. to N.E. Pt. of Texada I. including Howe Sound and Jervis Inlet (Strait of Georgia sheet 1) | Richards 1860 | 1863–1951 |
| 1863 Feb 23 | 585 | Harbours in Strait of Georgia: Port Augusta [*deleted 1901*]; Pender Harbour; Port Graves; Nanoose Harbour; Shoal Channel & Plumper Cove (Howe Sound) | Richards *Plumper* 1860 | 1863–1913 |
| | | [*later inset*] Union Wharf (Baynes Sound) [*deleted 1901*] | Masters 1893 | |
| 1863 Feb 28 | 2067 | Vancouver Island: Harbours in Discovery Passage, Broughton Strait, and Goletas Channel: | Richards *Plumper* 1860 | 1863–1954 |
| | | Bull Harbour | | |
| | | Beaver Harbour | [Simpson *Egeria* 1902] | |
| | | Gowlland Harbour [Duncan Bay] & Quathiaski Cove [*deleted 1904*] | | |
| | | Shucartie [Shushartie] Bay | | |
| | | Alert Bay and Entrance of Nimpkish River [*deleted 1904*] | | |
| | | Beaver Cove [*deleted 1904*] | | |

Table 3.1 (continued)

| *Published* | *Chart* | *Title* | *Surveyor and date* | *Current* |
|---|---|---|---|---|
| 1863 Jul 3 | 584 | Sydney Inlet to Nitinat including Clayoquot & Barclay [Barkley] Sounds | Richards 1861 | 1863–1942 |
| | | [*insets*] Refuge Cove; Uchucklesit Harbour; Island Harbour | | |
| | | [*later insets*] Stamp Harbour | Richards 1861; Walbran 1892 | |
| | | Entrance Harbour | | |
| 1864 Apr 16 | 572a | Naval Reserve (Duntze Head) | Richards *Hecate* 1863 | 1864–1870s |
| 1864 May 14 | 572 | Constance Cove (Esquimalt Harbour) | Richards *Plumper* 1858 | 1864–1923 |
| 1864 May 23 | 573 | Nanaimo Harbour | Richards *Hecate* 1862 | 1864–1901 |
| 1864 Jun 20 | 2627 | Semiahmoo Bay and Drayton Harbour | Richards 1857 | 1864–1958 |
| 1864 Oct 15 | 576 | Esquimalt & Victoria Harbours | Richards *Hecate* 1861–62 | 1864–1925 |
| 1864 Oct 15 | 577 | Inner Channels leading from Juan de Fuca Strait to Haro Strait | Richards *Hecate* 1862 | 1864–1928 |
| 1864 Dec 8 | 582 | Goletas Chl. to Quatsino Sound including Scott Islands | Richards 1860 | 1864–1950 |
| 1865 Mar 31 | 570 | Quatsino Sound | Richards *Plumper* 1862 | 1865–1927 |
| | | [*insets*] Hecate Cove; Koprino Harbour | | |
| 1865 May 30 | 590 | Klaskino and Klaskish Inlets and Anchorages (Brooks Bay) | Richards *Hecate* 1862 | 1865–1942 |
| 1865 Jun 19 | 1916 | Nootka Sound | Richards 1862 | 1865–1942 |
| | | [*inset*] Friendly Cove | Richards 1863 (Belcher 1839) | |
| 1865 Jun 27 | 589 | Esperanza & Nuchatlitz Inlets | Richards *Hecate* 1862 | 1865–1942 |
| | | [*inset*] Queen's Cove | | |
| 1865 Jun 29 | 569 | Esperanza to Clayoquot including Nootka Sound | Richards 1862 | 1865–1942 |
| | | [*inset*] Hesquiat Harbour | | |
| 1865 Jun 30 | 583 | Quatsino to Esperanza including Kyuquot Sound | Richards 1863 | 1865–1942 |

Table 3.1 (continued)

| *Published* | *Chart* | *Title* | *Surveyor and date* | *Current* |
|---|---|---|---|---|
| 1865 Jul 13 | 2431 | Cordova Bay [Port Simpson] to Cross Sound including Koloschensk Archipelago<br>[*inset*] Anchorage off Point Highfield [*deleted 1901*]<br>[*later insets*] Point Highfield Anchorage<br><br>Port Tongass | Vancouver 1792; Russian chart 1853 [U.S. charts to 1892]<br>Simpson *Devastation* 1862<br>U.S. Govt. Survey 1896 (Simpson 1862)<br>U.S. Govt. Survey 1883 | 1865–1932 |
| 1865 Aug 10 | 364 | Duncan Bay & Metlah-Catlah Bay | Richards *Hecate* 1862 | 1865–1913 |
| 1865 Aug 30 | 611 | Griffin Bay and adjacent Anchorages (Haro Archipelago) | Richards *Plumper* 1858 | 1865–1913 |
| 1865 Sep 25 | 1917 | Vancouver Island and adjacent shores of British Columbia | Richards 1859–65 (Kellett 1847; U.S. Coast Survey) | 1865–1956 |
| 1865 Sep 30 | 592 | Barclay [Barkley] Sound | Richards 1861 | 1865–1942 |
| 1865 Sep 30 | 602 | Roche Harbour and its Approaches (San Juan Island) | Richards *Plumper* 1857 | 1865–1902 |
| 1865 Nov 20 | 714 | Oyster and Telegraph Harbours (Stuart Channel)<br>[*insets*] Maple Bay; Osborn Bay | Richards *Plumper* 1859 | 1865–1907 |
| 1865 Dec 23 | 630 | Port Neville (Johnstone Strait)<br>[*inset*] Forward Harbour | Richards *Plumper* 1860<br>Pender 1865 | 1865–1932 |
| 1866 Jan 8 | 634 | Port Harvey (Johnstone Strait) | Richards *Plumper* 1860 | 1866–1912 |
| 1866 Jan 20 | 1907 | Sooke Inlet | Pender 1864 | 1866–1949 |
| 1866 Mar 24 | 716 | Nasparti and Ou-ou-kinsh Inlets | Richards 1863 | 1866–1942 |
| 1866 Mar 24 | 717 | Kyuquot Sound | Richards 1863 | 1866–1942 |
| 1866 Apr 18 | 571 | Harbours in Vicinity of Queen Charlotte Sound: Blunden Harbour; Cypress Harbour; Tracey Harbour; Cullen Harbour | Pender 1863 | 1866–1962 |
| 1866 Nov 12 | 538 | Seymour Narrows (Discovery Passage) | Richards *Plumper* 1860 | 1866–1896 |

Table 3.1 (continued)

| *Published* | *Chart* | *Title* | *Surveyor and date* | *Current* |
|---|---|---|---|---|
| 1867 Jun 20 | 581 | Johnstone & Broughton Straits and Queen Charlotte Sound with Knight Inlet & Adjacent Channels | Richards 1860; Pender 1863–65 | 1867–1935 |
| | | [*insets*] Sunday Harbour; Dusky Cove; Farewell Harbour | Pender 1865 | |
| | | [*later insets*] Su Quash Anchorage | 1869 | |
| | | Vere Cove | Watson/Smyth *Egeria* 1898 | |
| 1867 Dec 12 | 1923a | Cape Caution to Port Simpson including Hecate Strait and part of Queen Charlotte Islands [northern sheet] | Pender 1867 (Vancouver) [Parry, Learmonth *Egeria* 1907–09] | 1867–1932 |
| | | [*later insets*] Khutze Anchorage | Anderson/Walbran 1902 | |
| | | Kemano Bay | Hodgson *Shearwater* 1905 | |
| | | Kitlup Anchorage | Hodgson 1906; Learmonth *Egeria* 1907 | |
| 1867 Dec 12 | 1923b | [Cape Caution to Port Simpson including Hecate Strait and part of Queen Charlotte Islands] [southern sheet] | [Pender 1867 (Vancouver) [Parry, Learmonth *Egeria* 1907–09]] | 1867–1932 |
| | | [*inset*] Kimsquit | | |
| 1868 Feb 28 | 1901 | Ogden Channel with anchorages between Cape Caution and Port Simpson: Ogden Channel and adjacent passages; Alpha Bay; Stuart Anchorage; Holmes Bay; McLaughlin Bay; Carter Bay; Kynumpt Harbour; Goldstream Harbour; Namu Harbour; Safety Cove; Schooner Retreat | Pender 1867 | 1816–1962 |
| 1869 Nov 15 | 555 | Goletas Channel with passages leading into Queen Charlotte Sound | Richards *Plumper/Hecate* 1860–63 | 1869–1904 |
| 1872 Mar 11 | 2426 | Port Simpson & adjacent anchorages | Pender 1868 | 1872–1907 |
| 1872 Apr 24 | 2448 | Approaches to Fitz-Hugh & Smith Sounds | Pender 1870 | 1872–1891 |
| | | [*inset*] Takush Harbour | | |
| 1872 Apr 30 | 48 | Skidegate Inlet | Pender 1866 | 1872–1958 |
| 1872 Oct 1 | 2449 | Lama Passage and Seaforth Channel | Pender 1866–69 | 1872–1958 |

Table 3.1 (continued)

| *Published* | *Chart* | *Title* | *Surveyor and date* | *Current* |
|---|---|---|---|---|
| 1872 Oct 18 | 2453 | Brown & Edye Passages (Chatham Sound) | Pender 1867–69 [Dodge, Musgrave 1906–07; Parry *Egeria* 1908–09] | 1872–1913 |
| | | [*insets*] Qlawdzeet Anchorage<br>Refuge Bay | [Parry *Egeria* 1908]<br>[Nares/Parry *Egeria* 1909] | |
| 1872 Oct 25 | 2190 | Nass Bay | Pender 1868 | 1872–1954 |
| 1872 Dec 16 | 1462 | Anchorages adjacent to Fitz-Hugh & Milbank Sounds: Klemtoo Passage & Anchorage; Port Blakeney, Morris Bay & adjacent channels; Nowish Cove; Welcome Harbour; Bela-Kula Anchorage | Pender 1867–70 | 1872–1954 |
| 1872 Dec 16 | 2189 | Ports adjacent to Principe and Grenville Channels: Port Canaveral; Port Stephens; Lowe Inlet; Klewnuggit Inlet; Coghlan Anchorage | Pender 1868–70 | 1872–1950 |
| 1878 Jan 10 | 1906 | Becher and Pedder Bays | Pender 1870 | 1878–1949 |
| 1880 Nov 5 | 2430 | Queen Charlotte Islands and adjacent coasts of British Columbia | Pender 1867–70; Dawson 1879 | 1880–1914 |
| 1882 Aug 22 | 2689 | Haro and Rosario Straits [later: Juan de Fuca Strait to Strait of Georgia] | Richards *Plumper* 1858–59 (Kellett 1847); U.S. Surveys to 1856 | 1882–1952 |
| 1883 Jul 12 | 1911 | Juan de Fuca Strait<br>[*insets*] Port San Juan; Neeah Bay | Admiralty Surveys to 1883 | 1883–1900 |
| 1883 Jul 18 | 1947 | Admiralty Inlet and Puget Sound | U.S. Coast Surveys to 1880 | 1883–1893 |
| 1886 Jul 9 | 922 | Burrard Inlet | Richards *Plumper* 1859 | 1886–1893 |
| 1891 Nov 5 | 2448 | Approaches to Fitz Hugh & Smith Sounds<br>[*inset*] Takush Harbour | Pender 1870 | 1891–1952 |
| 1893 Mar 6 | 922 | Burrard Inlet<br>[*inset*] Vancouver Harbour<br>[*later insets*] First Narrows; Second Narrows | Stewart/Boulton 1891<br><br>Smyth *Egeria* 1899 | 1893–1929 |
| 1893 Sep 27 | 1792 | Port Townsend and Kilisut Harbour | U.S. Coast Survey 1893 | 1893–1995 |

Table 3.1 (continued)

| *Published* | *Chart* | *Title* | *Surveyor and date* | *Current* |
|---|---|---|---|---|
| 1893 Oct 9 | 1947 | Admiralty Inlet and Puget Sound | U.S. Coast Surveys to 1889 | 1893– |
| 1896 Apr 16 | 538 | Anchorages on East Coast of Vancouver Island: | | 1896–1928 |
| | | Menzies Bay | Chambers *Nymphe* 1895 | |
| | | Seymour Narrows | Richards *Plumper* 1860 | |
| 1896 Dec 15 | 2458 | Port Simpson to Port McArthur including Inner Channels and Prince of Wales Island | Latest U.S. & British surveys | 1896– |
| 1897 Dec 31 | 1717 | Port Angeles | U.S. Coast Survey 1893 | 1897–1993 |
| 1898 Sep 30 | 1835 | Clayoquot Sound | Richards *Hecate* 1861; Walbran | 1898–1942 |
| 1899 Feb 28 | 3029 | Active Pass to Gabriola Pass and Inner Channels | Richards *Plumper* 1858–59 [Walbran *Quadra* 1897–98] | 1899–1907 |
| | | [*inset*] Chemainus Bay | | |
| | | [*later inset*] Porlier Pass | Walbran *Quadra* 1900 | |
| 1900 Apr 14 | 2078 | Harbours and Anchorages in Strait of Georgia: | | 1900–1912 |
| | | Sturt Bay and Van Anda Cove | Walbran *Quadra* 1899 | |
| 1900 Apr 30 | 1911 | Approach to Juan de Fuca Strait | Admiralty Surveys to 1883; U.S. Govt. Chart to 1899 | 1900–1952 |
| | | [*insets*] Port San Juan (Port Renfrew); Neeah Bay | | |
| 1900 Jul 2 | 333 | Baynes Sound and Approaches | Smyth *Egeria* 1898 | 1900–1974 |
| 1900 Jul 26 | 3127 | Port Augusta (Comox) | Somerville/Smyth *Egeria* 1898 | 1900–1950 |
| 1900 Dec 21 | 2870 | Toba, Bute & Loughborough Inlets and adjacent channels [*insets*] Waddington Harbour; Beaver Creek; Gorge Harbour; Cameleon Harbr.; Drew Harbour; Squirrel Cove | Richards 1860–62; Pender 1864–69 | 1900–1929 |
| 1901 Mar 29 | 573 | Nanaimo Harbour | Smyth *Egeria* 1899 | 1901–1994 |
| 1902 Jan 25 | 3178 | Plans in Discovery Passage: Gowlland Harbour; Otter Cove; Elk Bay; Duncan Bay | Simpson *Egeria* 1900 | 1902–1961 |

Table 3.1 (continued)

| *Published* | *Chart* | *Title* | *Surveyor and date* | *Current* |
|---|---|---|---|---|
| 1902 Feb 5 | 3162 | Discovery Passage | Simpson *Egeria* 1900 | 1902–1997 |
| 1902 May 2 | 3271 | Plans in Broughton and Johnstone Straits:<br>Alert Bay<br>Blinkinsop Bay | Simpson *Egeria* 1900<br>(Richards *Plumper* 1860) | 1902–1954 |
| 1902 Jul 31 | 3260 | Johnstone Strait, sheet 1 (East) | Simpson *Egeria* 1900 | 1902–1997 |
| 1902 Nov 21 | 602 | Roche Harbour and its approaches (San Juan Island) | U.S. Govt Surveys 1894–95 | 1902–1958 |
| 1903 May 9 | 3333 | Johnstone Strait, sheet 2 (Central) | Simpson *Egeria* 1900–01 | 1903–1997 |
| 1904 Mar 11 | 3387 | Johnstone Strait, sheet 3 (West)<br>[*inset*] Forward Bay | Simpson *Egeria* 1901<br>Alexander 1901 | 1904–1997 |
| 1904 July 30 | 3417 | Hanson Island to Beaver Harbour including Broughton Strait<br>[*inset*] Port McNeill | Simpson *Egeria* 1901–02<br>Lay, Harris 1901 | 1904–1955 |
| 1904 Oct 8 | 3430 | Goletas and New Channels | Simpson, Parry *Egeria* 1902 (Richards *Plumper* 1860) | 1904–1955 |
| 1904 Oct 8 | 3443 | Nahwhitti Bar with passages leading into Queen Charlotte Sound<br>[*inset*] Bull Harbour | Simpson *Egeria* 1902 (Richards *Plumper, Hecate* 1860–62)<br>Simpson *Egeria* 1902 | 1904–1955 |
| 1904 Oct 8 | 3448 | Blunden Harbour | Edgell, Brandon/Parry *Egeria* 1903 | 1904–1953 |
| 1904 Oct 28 | 3447 | Moresby Passage with its approaches | Simpson *Egeria* 1902 [Parry *Egeria* 1905, 1908] | 1904–1938 |
| 1904 Nov 18 | 3462 | Anchorages in Queen Charlotte Sound:<br>Southgate Group and Anchorage | Parry *Egeria* 1903 | 1904–1953 |
| 1905 Oct 28 | 3520 | Active Pass | Parry *Egeria* 1904 | 1905–1994 |
| 1905 Nov 13 | 2512 | Approaches to Nanaimo Harbour | Parry *Egeria* 1904 (Smyth *Egeria* 1899) | 1905–1949 |

Table 3.1 (continued)

| *Published* | *Chart* | *Title* | *Surveyor and date* | *Current* |
|---|---|---|---|---|
| 1905 Dec 27 | 3517 | Nanoose Harbour and approach | Parry *Egeria* 1903–04 | 1905–1949 |
| 1905 Dec 27 | 3523 | Plans on East Coast of Vancouver Island: | | 1908–1961 |
| | | Oyster Harbour | Parry *Egeria* 1904 | |
| | | Dodd and False Narrows | Parry *Egeria* 1904 | |
| | | [*inset*] Dodd Narrows | | |
| | | [*later inset*] Chemainus Bay | Nankivell/Parry *Egeria* 1905 | |
| 1906 Dec 10 | 3594 | Plans on East Coast of Vancouver Island: | | 1906–1971 |
| | | Ganges Harbour and Captain Pass | Miles, Fraser/Parry *Egeria* 1905 | |
| | | Porlier Pass | Knight/Parry *Egeria* 1905 | |
| 1907 Jun 4 | 3618 | Moresby Passage to Gabriola Pass (Northern sheet) | Parry *Egeria* 1905 | 1907–1952 |
| 1907 Jul 30 | 3619 | Moresby Passage to Gabriola Pass (Southern sheet) | Parry *Egeria* 1905[-08] (Simpson *Egeria* 1902, Richards *Plumper* 1858–60) | 1907–1952 |
| 1907 Nov 16 | 3678 | Port Simpson | Learmonth *Egeria* 1906 | 1907–1955 |
| 1907 Dec 30 | 2426 | Port Simpson and adjacent anchorages | Learmonth *Egeria* 1906 (Pender 1868) | 1907–1961 |
| 1908 Nov 9 | 3713 | Masset Harbour (Graham Island) | Davy/Learmonth *Egeria* 1907 | 1908–1958 |
| 1909 Feb 9 | 3716 | Plans in Graham Island: Parry Passage; The Bar and Alexandra Narrows (Virago Sound); Virago Sound and Naden Harbour | Learmonth *Egeria* 1907 | 1909–1958 |
| 1910 Sep 29 | 3282 | Prince Rupert Harbour and Southern Approaches | Dodge, Musgrave, Parizeau, Davies (Canada) 1906–07 | 1910–1958 |
| 1911 Apr 11 | 3754 | Dixon Entrance | Learmonth, Parry *Egeria* 1907–09; Musgrave *Lillooet* 1909–10; U.S. Govt. chart 1910 | 1911–1992 |
| 1911 May 9 | 2840 | Haro and Rosario Straits | British & U.S. Govt. Surveys to 1908 | 1911–1995 |

# 4

# "The Incarnation of Energy": Raymond Préfontaine, the Hydrographic Survey of Canada, and the Establishment of a Canadian Naval Militia

RICHARD H. GIMBLETT

The general experience of most Western navies has been for a hydrographic survey to be established as a logical adjunct to a mature and thriving naval service. That would make Canada perhaps unique in having established the Hydrographic Survey of Canada (later the Canadian Hydrographic Service, or CHS) in 1904, several years before the creation of the Royal Canadian Navy (RCN) in 1910 (or much earlier, if one looks to the Georgian Bay Survey of 1883). Indeed, the establishment of the CHS was an important step towards the formation of a Canadian naval service separate from the Royal Navy. We tend to forget that much of the Canadian "nation-building" at the turn of the last century was undertaken in the context of defining Canada's place as a self-governing dominion within the British Empire. Among the forces driving the establishment of the CHS were not only the growing sense of autonomy but also the notionally un-Canadian spirit of navalism that otherwise infused the first decade of the twentieth century. Few politicians have attempted to mark their rise by building a navy – Teddy Roosevelt, who became president of the United States in September 1901, was one. What makes the situation of the Hydrographic Survey of Canada all the more revealing is that its promoter – the minister of Marine and Fisheries, the Hon. Raymond Préfontaine (who bore a passing physical resemblance to President Roosevelt) – was a French Canadian in Sir Wilfrid Laurier's cabinet, and he saw the CHS as part of the logical progression towards a Canadian naval militia.

Joseph-Raymond Fournier Préfontaine is but one of those many Canadian ministers of the Crown who have been overlooked by history, in his case primarily because of his premature death from a heart attack on Christmas night, 1905, at the age of fifty-five. Springing from modest habitant stock, he would not seem to have been destined for a role as a navalist (one more commonly associated with English Canadians), but in many ways he was a natural. Préfontaine craved power, and in that age especially, "navy" meant power – not only in its purest military form but also in terms of personal prestige and opportunities to dispense patronage.[1] At the time of his death, Préfontaine's naval-based power was building to the point where Laurier was beginning to perceive a threat to his leadership of the Liberal Party.

Préfontaine's resumé reads like the model for a perennially ambitious hustler. Having left the family farm near Longueuil to study law at McGill University, he articled in an office of "dyed-in-the-wool Rouges," married profitably into a family of *Bleu* merchants, and upon being called to the bar, joined an established English Montreal firm.[2] With that impeccable range of connections, he promptly embarked upon a career in politics, at one point holding office at all three levels – municipal, provincial, and federal – at the same time. An early assessment summed up his priorities neatly: he was "a good lawyer" but a "mediocre clerk" because "boring procedural details were not to his liking and he neglected the study of the profession in favour of battles on the hustings."[3]

A full-length biography of Préfontaine has yet to be undertaken, and those studies that do include him focus mostly on his career in Montreal municipal politics, with only passing reference to his truncated career at the federal level. Indeed, it was in civic politics that he established his modus operandi. One biographical account notes, "There was something of the visionary about Préfontaine." "He was a man of action whose strength ... lay in the 'rapidity of his judgement,' his genius for organizing, and his capacity for work." But his "intense activity raised questions."[4] Invariably, his sense for the efficiencies to be gained through consolidation of administration would succumb to the temptations of patronage politics. As mayor of the village of Hochelaga, he led the drive for its incorporation into the city of Montreal in 1883, after which he became an alderman in the greatly expanded metropolis. Appointed chairman of the Roads Committee just as the demand for road construction and paving increased to make that the fastest growing segment of the city's expenditures, Préfontaine immediately appreciated this was an ideal position from which to fashion a formidable political machine.[5] In consequence, he dominated Montreal municipal politics from 1886 to 1902.[6] He liked to style himself "the voice both of the east-end and of progress,"[7] and in truth he was instrumental in raising the influence of the lower-class French-Canadian majority in city issues. The irregularities in his backroom methods, however, served to inspire the counter-development of an urban reform movement through the latter 1890s. Although the reformers were able to overthrow his machine in the municipal election of 1900, his own re-election that year for a second consecutive term as mayor (in a city where tradition dictated alternating between French and English incumbents serving one term only) underscored his personal popularity with a broad cross-section of voters.[8] Still, with this net decline in local influence, he began to shift his focus to the national level.

Préfontaine was first elected federally in the July 1886 by-election in Chambly, running for the Liberals to protest the execution of Louis Riel, and he remained a fixture of the Liberal caucus until his death, changing seats only to represent the new riding of Maisonneuve (formerly Hochelaga) in 1896, in the pivotal election that swept Wilfrid Laurier to power. As an *ancien* of the Liberal Party and still holding his influential municipal position, Préfontaine fully expected to be appointed Laurier's lieutenant for the province of Quebec. That honour, along with the ministry of Public Works – noted elsewhere as "an office befitting Préfontaine's ambition"[9] – went instead to Joseph-Israël Tarte, as reward for his having crossed the floor from the Conservatives to join Laurier. Accordingly, Préfontaine continued his municipal activities, but his election in 1898 to the pinnacle of local power as mayor occurred even as Tarte's stature in the Liberal Party was beginning to wear thin. Although "his administration of his department was most effective,"[10] Tarte soon found his *bleu* instincts were not always compatible with Laurier's need to demonstrate political flexibility. A major clash came in the fall of 1899, when Tarte opposed the dispatch of a Canadian contingent to South Africa against the Boers. Perhaps to spite his nemesis, Préfontaine waded into the issue in favour of participation and was commended in the English press for representing "the real sentiments of French Canadians."[11] He evidently warmed to the issue, and ultimately he was credited with carrying Montreal for Laurier, to the lasting gratitude of the prime minister. Nor was the consequent development of a broader English-Canadian constituency of his own lost on Préfontaine.

Laurier's pleasure sprang less from the resolution of the contingency crisis in itself than from the calm that settled the national focus once again upon the development of the young dominion's infrastructure. At the beginning of the century that he soon would proclaim as belonging to Canada, Laurier's vision of achieving that goal varied only in degree from the two-pronged approach of Sir John A. Macdonald's National Policy: the exploitation of the west as the "bread-basket of the world" and the establishment of Ontario as the industrial heartland. Both objectives came together in fostering any number of major national projects: the building of additional northern railways to open new areas to farming and to establish alternative ports to Vancouver and the Lakehead at Prince Rupert and on Hudson Bay; the building of a new shipping canal from Georgian Bay to the Ottawa River to bypass potential American interference on the lower Great Lakes; and the upgrading of the St Lawrence shipping channel.

Tensions between Britain and the continental European powers during the Boer War served to renew interest in another dimension of infrastructure development – the security of imperial communications from Britain to Australia through Canada. An idea that eventually would be expressed as the "All-Red Route" (from the colouring of British colonies on contemporary maps) had received its original impetus with the completion of the Canadian Pacific Railway in 1885 and, indeed, had led to the establishment of the Canadian Pacific Steamship line, facilitated by massive British and Canadian government subsidies to promote British-owned mail steamers available for conversion in wartime into armed merchant cruisers. The weak link in the system existed, ironically enough, on the Atlantic portion. The old Canadian merchant shipping trade (based primarily in Halifax and Saint John, New Brunswick), which had dominated the North Atlantic through most of the nineteenth century, had been forgotten in the shift of focus inland. Neither had it kept pace with the coincident shifts from wind to steam and wood to steel, and it was verging on collapse. The lure of mail subsidies had not been enough to offset the competitive advantages of shipping proceeding direct to New York. Canadian companies, however, continued to hope for government projects to encourage the redevelopment of the major ports of Atlantic Canada. Where rail had dominated the focus of Macdonald's vision in the National Policy, the twentieth century brought with it a multiplicity of options, many involving marine transportation and not all of which could be afforded at once.

The variety of means to the common end practically ordained that another point of departure amongst the three French Canadians was not long in coming. The position of mayor of Montreal brought with it a seat on the Harbour Commissioners of the port, which at that time fell under the federal supervision of Tarte's Public Works department. When, in the spring of 1902, a number of Canadian business interests (including the Bank of Montreal and the Canadian Pacific Railway, both represented by Lord Strathcona) attempted to revive the long-dormant issue of a fast Atlantic steamship service, there was general agreement that the winter terminus should be Halifax. As for a summer terminus, Préfontaine naturally put forward the candidacy of Montreal, seeing this as a great opportunity to subsidize the redevelopment of that port's antiquated facilities, and he had some success in swinging Strathcona's support in that direction. Tarte, meanwhile, felt that it should be Quebec City, close to his own riding.[12] Whatever the merits of either proposal, the import of the whole issue was raised considerably when Tarte opted to tie his advocacy of the project to a strengthening of a Canadian protectionist tariff, in memory of Macdonald's National Policy – and in direct contradiction to Laurier's preference for freer trade, both with the United States and within the empire.

Laurier was in England that summer for the Colonial Conference and other affairs of state, but immediately upon his return on 20 October 1902, he forced Tarte to resign.[13]

For all his political resiliency and although only sixty-one years old, Wilfrid Laurier was physically ailing, and he seriously considered resigning that fall. The battle with Tarte had in the process unleashed a host of other problems revolving around preferential tariff, marine transportation, revival of the shipbuilding industry, and ministerial responsibility. But it also seemed to revive the prime minister, and the three full weeks that Laurier took to organize the cabinet shuffle were put to ushering in a major revision to the machinery of his government.[14] Although the Public Works ministry that many had seen as Préfontaine's natural element now was open, the priorities for the development of a national infrastructure were shifting to the marine sector: "the completion of a Canadian system of both inland and ocean transportation" was beginning to figure as "the greatest of Imperial questions just as it is the greatest of national questions."[15] The cabinet shuffle brought with it the intended transfer of responsibility from the care of Public Works to that of Marine and Fisheries for all works and improvements in the St Lawrence River and for the building and maintenance of ports throughout the dominion. Previously, Marine and Fisheries had been a sleepy backwater of buoys and fish, and the incumbent, James Sutherland, had been an uninspired minister. For the task at hand, Laurier needed "a strong, aggressive man with many qualities of leadership ... [who was] the incarnation of energy"; and that is just how the Liberal Party organ, the Toronto *Globe*, heralded Préfontaine's appointment as minister of Marine and Fisheries on 11 November 1902.[16]

Préfontaine wasted little time settling into his new responsibilities. On 26 November he published a letter in the Montreal press "outlining his proposed policy in the Marine and Fisheries Department." Obviously addressed to his immediate constituents, it contained as a major element a promise "to complete the equipment of the Port of Montreal 'in the most efficient and permanent manner.'"[17] But already Préfontaine was recognizing opportunities to broaden his horizons. At a banquet in Montreal on 19 November, in honour of the general officer commanding the Canadian militia, British Major-General the Earl of Dundonald,

Préfontaine announced that he "hoped the time would soon come when Canada would organize at least the nucleus of a navy and believed that if Parliament took such a step it would meet with the endorsation of all Canadians."[18] To have made such a pronouncement to such an audience and so soon after his appointment was a bold move, even for the ambitious Préfontaine. At the height of the Pax Britannica, the Royal Navy was the unchallenged "Mistress of the Seas," and the need for separate colonial forces was hardly apparent, especially to a neophyte such as Préfontaine. Almost certainly, the notion had to have been authorized – if not initiated – by Laurier himself.

On that score, it was hardly a new concept. Provision for a naval militia had existed in the dominion statues since the *Militia Act* of 1868, and even though one had never been instituted, the idea had enjoyed the highest level of official notice on several occasions. Sir John A. Macdonald had sanctioned the development of just such plans while prime minister,[19] and Laurier himself had been made to consider it. Indeed, the first broad-based public discussion of the issue had appeared in the *Globe*, courtesy of his prominent Toronto supporters (who happened also to belong to the Navy League of Canada), literally on the eve of his election victory in 1896.[20] Since then, Laurier had come to appreciate that Britain's growing rapprochement with the United States was making it awkward for the Royal Navy to intervene on behalf of Canada in a variety of maritime disputes, ranging from ongoing fisheries issues through the intrusion of American whalers into Hudson Bay to the Alaska boundary dispute.[21] The only alternative enforcement agency was the Canadian Fisheries Protection Service, but its small fleet of mostly steam-assisted wooden schooners could project only so much influence. Something more robust was required.

Necessity being the proverbial mother of invention, a confluence of needs had occurred at the Colonial Conference that summer of 1902. Over the years, there had developed a standard theme at these events, of the Admiralty requesting contributions from the colonies to the upkeep of the Royal Navy. When Colonial Secretary Joseph Chamberlain raised the issue this time, knowing Laurier's famous previous quip, "If you want our aid, call us to your council," he pre-empted its repetition with the following statement, which is worth repeating at length for its weighty significance:

> Gentlemen, we do want your aid. We do require your assistance in the administration of the vast Empire which is yours as well as ours. The weary Titan staggers under the too vast orb of its fate. We have borne the burden for many years. We think it is time that our children should assist us to support it, and whenever you make the request to us, be sure that we shall hasten gladly to call you to our Councils. If you are prepared at any time to take any share, any proportionate share, in the burdens of the Empire, we are prepared to meet you with any proposal for giving to you a corresponding voice in the policy of the Empire.[22]

Complicating the issue, when the specific subject of imperial naval defence arose during the second session, the first lord of the Admiralty, the Earl of Selborne, introduced a major shift in naval strategy. Rather than previous standard notions of defence, henceforward the only method to protect the commerce and territory of the empire would be to adopt an offensive posture of concentrating "the greatest possible force where [the enemy's] ships are, and to destroy those ships," even if it followed that there "could be no local allocations of ships to protect the mouth of the Thames, to protect Liverpool, to protect Sydney, to protect Halifax."[23] However marginally effective Laurier might have perceived the British squadrons at Esquimalt and Halifax, a complete withdrawal was sure to raise an outcry from the local citizenry. Canada's interests were best served by maintaining the status quo. He and his attending ministers quickly drafted a "Memorandum by the Canadian Ministers Concerning Defence," in which the message on the naval question was quite clear: while Canada was firmly against a policy of contribution, that was because the dominion also "[valued] highly the measure of local independence which has been granted it from time to time by the Imperial authorities"; to that extent, even though at present "Canadian expenditures for defence services are confined to the military side," if (it was implied) the naval defence of Canadian coasts could not be guaranteed by the Royal Navy, then the "Canadian Government are prepared to consider the naval side of defence as well."[24]

But the navy almost by definition was the most imperial of institutions, and Laurier was careful to hedge his bets. In other sessions, he had already emphasized that he considered expenditures on public works such as "canals, railways, harbours, improvements in rivers and so on" to be as important as and probably proportionately equitable to Britain's own military expenditures (noting also that Britain did not have to invest in such public works).[25]

It was in this frame of mind that Laurier returned to Canada, determined to continue the priority development of national transportation infrastructure (but now especially the marine sector) and counting those expenditures as an offset to imperial contributions, but also newly committed to the development of a naval policy and having to undertake the sacking of Israël Tarte. The promotion of the resourceful Raymond Préfontaine to the Marine and Fisheries portfolio seemed to solve a variety of problems at one stroke. Marine infrastructure would continue to receive the attention of an interested and energetic minister. It was a definite bonus having a French Canadian who had represented the government's interests so well on previous imperial issues (the Boer War contingents) available now to act as a sounding board on the idea of a Canadian naval service.

In that respect, Préfontaine's Montreal speech in November attracted no negative attention. Later that month a senior official of the British Navy League, H.F. Wyatt, arrived to capitalize on the interest in naval affairs generated at the colonial conference. Canadian branches already existed in Toronto (established 1895) and Victoria (1901). Wyatt's two-month cross-country tour served to invigorate those branches and saw the organizing of new ones in practically every other major urban centre. Invariably, the meetings would result in the unanimous passing of a motion along the lines of that put forward at Saint John, New Brunswick: "This meeting approves of the proposed creation by the Dominion Government of a Canadian naval force, and expresses the earnest hope that in its future development the principle of Canadian autonomy may be reconciled with that of the strategic unity of the Imperial Navy and of the integrity of the latter as a single force." The non-partisan nature of this support was given some semi-official impetus when a variety of government ministers chose to appear among the list of prominent citizens named to honorary positions, including Préfontaine as president of the Ottawa branch.[26]

Priorities, however, were priorities, and Préfontaine, as promised, turned his "genius for organizing" and "capacity for work" to the task of rationalizing the interdepartmental responsibilities for marine matters. These turned out to be vaster and of greater consequence than perhaps either Laurier or Préfontaine had envisioned. Indeed, the new minister's attention became focused only a few days after his "nucleus of a navy" speech before Dundonald, when, on the morning of 24 November 1902, shortly after departing Quebec City for England, the Allan liner SS *Sicilian* "in the middle of the channel ... struck some object, either a sunken wreck, or a huge boulder."[27] Despite flooding in the forward hold, the ship was able to return to Quebec, effect repairs, and depart before the winter freeze-up. Even though the subsequent inquiry by the Wreck Commission could determine no blame to the pilotage or any other authorities,[28] there were connections to be made. This grounding had occurred in the same place that the SS *Lake Huron* "sustained considerable damage some years ago. It is also claimed that no survey of the locality has been made since."[29] When the Lloyd's correspondent for the London *Times* went so far as to "doubt whether the special favourable terms now granted the Allan Line are adequate in view of recent disasters," the Canadian government promised immediately to undertake to sweep the St Lawrence channel the next season and to prepare a new survey.[30]

It should be noted that the two previous major dominion hydrographic surveys had been initiated as the direct consequences of marine disasters for which the failure of the Canadian government to have provided accurate charts made it at least partly to blame: the Georgian Bay Survey had commenced in 1883 after the loss of the steamer *Asia* in September 1882,[31] and the Burrard Inlet Survey was completed in 1891 after the Canadian Pacific liner *Parthia* "encountered a shoal" there in June 1890.[32] If Laurier and Préfontaine had needed any encouragement not to delay their announced amalgamation, this latest brush with disaster provided fair warning. The viability of many of the envisioned marine transportation infrastructure projects ultimately rested upon reliable charting. To varying degrees, the fortunes of harbour and port improvement, the Georgian Bay Shipping Canal, the northern rail terminus at Prince Rupert, the Hudson Bay Route, the St Lawrence Shipping Channel, and the fast Atlantic steamship line all hung upon it.

The broad thrust of the major shifts from Public Works to Marine and Fisheries were sketched out in rough drafts of the appropriate legislation by the end of the 1902–03 fiscal year:[33] the management and control of the St Lawrence Shipping Channel and general navigation affairs, including all dredging and sweeping equipment, together with control of the harbour commissions of the country. But putting it all into effect entailed a variety of smaller items, including the one of significance to readers of this volume: the amalgamation of the hydrographic work of the departments of Public Works and of Railways and Canals with that of Marine and Fisheries into one Hydrographic Survey of Canada. The Order-in-Council (PC 461) containing the authority for the establishment, as signed by the president of the Privy Council, Prime Minister Laurier, and approved by the governor general, the Earl of Minto, on 11 March 1904 (the centennial commemorated by this volume) affords an appreciation of the level of detail required:

> (i) ... with a view of systematizing and facilitating the work in connection with Hydrographic Surveys, the administration of which branch of the public service is assigned to the Department of Marine and Fisheries under the provisions of 55–56 Vict. c. 17, and the work thereof has been continuously performed by the Department for many years past, that all the Hydrographic work of the Departments of Public Works and Railways and Canals be transferred to the Department of Marine and Fisheries, and that Department alone be charged in future with the management and control of such surveys.
> (ii) ... that any records and plans in the possession of the Department of Public Works or of Railways and Canals ... be transferred to the last named Department, upon its asking applications for ...
> (iii) ... that all moneys voted by Parliament to either the Department of Public Works or that of Railways and Canals ... be placed to the credit of the last named Department.
> (iv) ... that the changes as recommended above take effect from this date.[34]

An additional, complementary Order-in-Council (PC 1200), dated 4 August 1904, appointed an experienced Canadian hydrographic surveyor to head the amalgamated survey:

> all hydrographic work in the Dominion should be under the management and control of a Chief Hydrographic Surveyor having his headquarters permanently at Ottawa ... It will therefore be seen that Mr. William J. Stewart has a very large and long experience as a Hydrographic Surveyor upon the Great Lakes, and has also some salt-water experience where currents were very strong and tides of great range ... The Minister therefore recommends that Mr. Stewart be appointed Chief Hydrographic Surveyor of Canada.[35]

Altogether, because of the span and intricacy of the rearrangements, the whole process would not be announced until 22 April 1904. At that time, they were heralded as "An important change in the internal and administrative composition of the Government"; indeed, the actual budgetary expenditures involved were so large that it was stated, "These changes ... put Mr. Préfontaine in control of the greatest spending Department of the Dominion Government."[36]

The consolidation under Préfontaine's control of the Hydrographic as well as the Tidal and Current Surveys and the chain of newly constructed Marconi wireless telegraphy stations along the east coast was not undertaken out of any conscious concern for a future naval service. The greatly expanded role of the Department of Marine and Fisheries did have a profound influence on naval development, however, primarily in enlarging the bureaucracy of the department and adding to the professionalism of its officers. The example of the Hydrographic Survey is a case in point. Before 1903, the efforts in the other departments (Public Works and Railways and Canals) were irregular, and within Marine and Fisheries itself they consisted only of the officer-in-charge of the Georgian Bay Survey (William J. Stewart) and two assistant clerks. The amalgamation brought with it nineteen engineers, surveyors, and assistants from the other departments, and the broadened scope of work allowed for the hiring of an additional five assistants. The combined total of twenty-seven represented a ninefold expansion and allowed Stewart (who began to style himself the "Chief Hydrographer") to carry out his plans for future growth in an efficient manner.[37] Considering that he was effectively starting from nothing, Stewart made a most auspicious debut. The first Canadian chart produced from Canadian surveys was published in February 1903, although since it was for the "Southern Portion of Lake Winnipeg," the demand proved "exceedingly small."[38] A more useful and popular publication would be the public issue in July 1904 of the first Canadian chart of the Great Lakes, a preliminary, unnumbered photolithographic edition entitled "Victoria Island to Fort William [Thunder Bay], Lake Superior."[39]

This admittedly slight but useful progress indicates that the instincts driving the amalgamation were in a sense a break from Canadian habit and proffered the image of a confident and maturing nation. Whereas the two previous "Canadian" surveys had both been undertaken under British Admiralty auspices, now the tables were turned. When, in May 1904, the Admiralty "of the weary Titan" issued a circular to the self-governing colonies requesting them "to conduct hydrographic surveys along their own coasts," Stewart could take some satisfaction in having anticipated the appeal. Still, for all of their good intentions, there was only so much that his limited number of surveyors could accomplish. The official response through the colonial secretary in London (made only in January 1905, an indication of the limited capacity of the Canadian bureaucracy) was forwarded under a minute by the new governor general, Earl Grey, "expressing concurrence in the desirability of Canada's undertaking the marine surveys of her own coasts" but requesting "that until arrangements for giving assent to this plan are completed ... the Admiralty will allow their surveying vessels on those coasts to continue their work there."[40] This request met with the satisfaction of the Admiralty, as the requirement on the east coast had already effectively ceased, but a Royal Navy vessel remained in British Columbia waters until the formal transfer of the Esquimalt Dockyard in November 1910.[41]

Préfontaine was coming to appreciate the vastly increased spending power of his department. In June 1903 he received tenders for the Fast Atlantic Line project, and while none of them in the end were deemed acceptable, within days the Montreal Harbour Commissioners were extended "a loan of $3,000,000 ... for the purposes of local improvement."[42] Otherwise, the "boring procedural details" of amalgamation continued to occupy the bulk of 1903 and, indeed, ran well into 1904. Préfontaine began to feel the itch to return to "battles on the hustings." The call was answered in a fashion tailor-made for his new position of power. Early in 1903, when the focus still was on the naval defence of the empire and the possible organization of a Canadian naval militia, only the Toronto *Star* drew the connection that "if Canadians desired Great Britain to stand by them to the point of war in such matters as that of Alaska they should aid in the organized defence of the Empire."[43] When the result of the Alaska boundary arbitration was announced on 21 October 1903, it was on terms unfavourable to the Canadian position and mostly due to the siding of the British representative on the commission with the Americans. The Canadian government – indeed, the general public – was quick to draw the implication that once again the imperial power had sacrificed Canadian interests in favour of rapprochement with the American republic. This response led almost immediately to calls for a revision to the treaty-making power (still legally resident with the imperial government in London), and Préfontaine was among the biggest boosters of the issue, appearing in public with Laurier on several occasions to make impassioned speeches in favour of greater Canadian autonomy in all respects.[44]

Préfontaine also was quick to realize the implication of the connection made earlier that year by the Toronto *Star*. While in Toronto in late November, speaking to the issue of the treaty-making power, he also took the opportunity to make a long-deferred announcement of the local construction of a new fisheries protection cruiser for the Great Lakes.[45] No ordinary fisheries vessel, the steel-hulled, screw-driven, ram-bowed Canadian Government Ship (CGS) *Vigilant* has been described as "the first modern warship to be built in Canada."[46] At about the same time, Préfontaine ordered an even more warlike vessel to upgrade the ranks of the Fisheries Protection Service on the east coast, where poaching by American fishermen was a constant annoyance.[47] That vessel, eventually to be commissioned as the CGS *Canada*, was ordered from the major British shipbuilding firm Vickers Sons and Maxim to an early pattern of ram-bowed torpedo-gunboat; armed with 3-pounder quick-firing guns, she was described at the time as a "third-class cruiser."[48] Upon her acceptance into service in the late fall of 1904, Préfontaine would proudly proclaim her interchangeably as the "flagship" or the "nucleus of the Canadian navy."[49] Indeed, at the time the two new vessels were ordered, an expenditure was placed in the departmental estimates for a "study into the establishment of a Canadian Naval militia."[50] Separate plans also were announced for the acquisition of a new hydrographic survey vessel for the British Columbia coast, obviously intended to demarcate the offshore boundary with Alaska.

Throughout 1904, as these various measures were introduced before Parliament, they were greeted with the enthusiastic support of the Opposition. When the minister of Militia and Defence, Sir Frederick Borden (a cousin of the Conservative leader), introduced his long-awaited bill to deal with militia reform, it generally was expected that a naval militia would be provided for in a separate bill. A draft *Act Constituting the Naval Militia of Canada* had indeed been prepared,[51] although it seems that even with the evident popularity of the notion, Laurier remained unsure of its potential reception in Quebec and for the moment was content with the status quo. The act was never placed before Parliament, to the curiosity of the Opposition, who charged that Préfontaine was pursuing the naval militia project in secret so as to hide the additional increase in his personal budgeting power.[52] The situation seemed to change in October 1904, when Admiral Sir Jackie Fisher was made first sea lord and immediately put into action the plan proposed at the Colonial Conference of 1902 for closure of the Royal Navy's foreign stations and concentration of the Home Fleet. But before Préfontaine could proceed, the British army suddenly announced its own intention to withdraw the imperial garrisons from the fortresses in Halifax and Esquimalt. Raising a force of dominion troops to replace them became the government's priority, but the large increase in strength of the permanent force led to an increase of nearly 40 per cent in the militia department's expenditure.[53] As the *Canadian Military Gazette* opined, "we cannot have everything, and that the assumption of these obligations will undoubtedly postpone the day when we may expect substantial Government assistance towards a navy."[54] More pointedly, the Toronto *Globe*, formerly a proponent of a naval militia, now sounded a cautionary note: "The surest defences of Canada are not in the multiplication of forts and guns and torpedo boats, but in a rapidly-increasing population, intelligent, prosperous, united and free."[55]

Préfontaine was undeterred. If he could not have his naval militia, he at least had his ships. The two new cruisers had been accepted into service in the fall of 1904, and the following February *Canada* was dispatched to the Caribbean for three months' training with the North America and West Indies squadron of the Royal Navy. These activities were proudly detailed in the department's annual report for 1905, which also recorded a visit by Préfontaine to Halifax to tour the *Canada*, where "the Honourable the Minister was received on board by a guard of honour, and after leaving the ship's side was saluted with 11 guns."[56] So encouraged, he commissioned an "official" history of his department, the broad thrust of which was to promote the naval militia project, claiming, as it did, "Few, if any, of the works undertaken by the present administration of the Dominion promise to be of greater national importance than the organization of a Naval Militia."[57] (As an aside, it is interesting to note that the author was a prolific chronicler of militia affairs – he would go on to write many regimental histories – but he evidently did not know enough of things nautical to make the Hydrographic Survey a major focus of the "history." Indeed, probably because the CHS remained a low spending priority, it was barely mentioned.) The hagiographic volume was published in November 1905, on the eve of a planned trip by Préfontaine to England to discuss various matters concerning his department with the appropriate imperial authorities. To Préfontaine's disappointment, the list, as approved by Laurier, did not contain any specific reference to a naval militia, dwelling instead on the routine of "pilotage matters [and] management of large shipping ports."[58]

The minister of Marine and Fisheries had a different idea of the purpose and priorities of his mission. Préfontaine used the few weeks ahead of his departure to promote his many varied interests, probably aided and abetted by distribution of complimentary presentation copies of the "history." The *Canadian Military Gazette* reported the purpose of this "most active and progressive Minister ... [was] to gather information to aid him in establishing the germ of a Canadian navy."[59] The editor of the Winnipeg *Free Press* (a faithful Liberal)[60] felt that such talks heralded a new stage in the constitutional development of the dominion: "Canada in taking over the entire responsibility of her own defence; in modernizing her militia system; in making a start in the establishment of a Canadian navy, makes it very clear that she intends to be a factor in future in the world politics in alliance with the Mother-land. But it will be an alliance, not a merger."[61] Other equally influential voices joined the refrain,[62] such that Laurier felt compelled to cable Préfontaine in London, "No arrangement should be undertaken with [the] Admiralty unless previous consultation with us here."[63]

In a letter to his deputy minister, Préfontaine professed not to understand the prime minister's attitude, feeling that he had exacted many useful concessions from the Admiralty, particularly "la cession par l'Amirauté anglaise des quais, constructions et bâtisses de Halifax et d'Esquimalt," as well as "des renseignements que l'on m'a fournis volontiers sur ce qu'il y aurait à faire pour organiser une milice navale."[64] He cabled Laurier himself from Paris the next day: "There is no reason to fear complications through the good work that has been done by Strathcona [in his capacity as high commissioner] and myself."[65]

It was to be Préfontaine's last communication with the prime minister. His death in his sleep four days later prevented a potentially ugly scene upon his return to Ottawa. Laurier did not object to the idea of a naval militia – it was, after all, government policy – but, rather, to the way that Préfontaine was going about it: using it to build his own popularly based "empire," and in a fashion reminiscent of Israël Tarte, as a possible challenge to Laurier.

Outwardly, however, there was no reason to be anything but magnanimous. Lord Tweedmouth, the first lord of the Admiralty, cabled Laurier expressing the regret of the imperial government, and making the battleship HMS *Dominion* (commanded by a Canadian in Royal Navy service, Captain Charles Kingsmill) available to convey Préfontaine's body to Canada. Laurier gratefully accepted the offer, and "The remains left Cherbourg for Canada on Jan[uary] 12th [1906], after impressive funeral services and international compliments in France."[66] Préfontaine was eulogized in the *Canadian Annual Review* for 1905 as follows:

> While party opinion was sharply divided as to the character and result of his earlier political life there was much general appreciation of his

administration of Departmental affairs and the press voiced a very wide feeling that his death removed a capable and valuable member of the Government as well as a popular political personality. His funeral, amid all the pomp and ceremony which the French and British Governments could accord, and with the final tribute of a British warship to convey his remains to Canada, marked the close of the year and the opening of a new one; as his successful negotiations with the Balfour Government for the acquisition of the Imperial Docks at Esquimalt and Halifax, and the establishment of a Canadian Naval Militia, had marked the closing days of an Imperial Ministry and distinguished the last days of his own life.[67]

Such lofty sentiment begs comparison with the actual record. The transfer of control of the imperial dockyards was perhaps Préfontaine's greatest achievement. Although Laurier hardly approved of the measure, he eventually sanctioned their acquisition to proceed – a significant act, for, as the official historian of the RCN observes, the implication was that the "ownership of bases suggests the advisability of owning warships as well."[68] And indeed, the Canadian government did own a "warship" already in the *Canada*, even if the fleet proper would not be established for another half-decade. Assessment of the administration of Préfontaine's department, however, would prove problematic. His body was barely cold before the Opposition began demanding an inquiry into the untendered fitting-out of the steamer *Arctic* to survey the Hudson Straits;[69] soon that would reveal an extensive patronage entanglement which would take his successor three years and a special royal commission (the Cassels Commission of 1908) to sort out. Yet, ironically, that process would prove the catalyst to complete the professionalizing of the department, so necessary for the establishment of a proper naval service.[70]

Out of it all, the one section of the Marine and Fisheries department that remained free from criticism was the Hydrographic Survey. Stewart evidently ran a tight but efficient ship. When the Royal Canadian Navy was authorized in May 1910, a Department of the Naval Service was established to oversee its administration. The same minister remained responsible for both departments, but again some rationalization of services was required. By Order-in-Council (PC 21–1453) dated 4 July 1910, the Hydrographic Survey of Canada was transferred to the new department, along with the Fisheries Protection Service, the Wireless Telegraph Branch, and the Tidal and Current Survey.[71] It would be some months (not until late fall) before the Canadian Navy's first true warships would arrive in Canada, on acquisition from the Royal Navy, thus confirming the distinctly Canadian order of seniority (in a fashion) of the Hydrographic Survey, a status unique amongst Western navies. Another form of Canadian distinctiveness would arise with the divorce of the CHS from the Naval Service and its return to the Department of Marine and Fisheries in 1922. But that is another story.

# 5

# *Hydrographic Survey Work of the Departments of Public Works, Railways and Canals, and the Interior, 1867–1914*

CHRISTOPHER ANDREAE

When Canada came into existence in 1867, an extensive water-based transport infrastructure was already in place. The former colonies had built canals, improved waterways, and erected a few lighthouses. The dominion hydrographic surveys undertaken during the first forty years of Canada's history expanded upon this base and greatly increased the scope of work. New projects were associated with a diverse range of navigation improvements, such as dredging, aids to navigation, and canal and harbour construction. The expansion of water transport facilities to accommodate an expected growth in traffic at the turn of the twentieth century precipitated a great increase in survey activity.

Modern academics accept without comment that all Canadian hydrographic services were consolidated into the Department of Marine and Fisheries (DMF) in 1904. While, indeed, the survey work of the Department of Public Works (DPW) and the Department of Railways and Canals (DRC) were transferred at that time, the full range of other hydrographic studies that had been and continued to be undertaken by various federal departments is ignored. An examination of that work is of interest for at least two reasons. First, it highlights the problems of interdepartmental jurisdiction that arose over time and that were dramatically emphasized by the grounding of the *Sicilian* in 1902. Second, a discussion of the overlap between engineering-related hydrography and the hydrography of navigation reveals the wide scope and technical complexity of work undertaken by the various government departments. Indeed, this diversity of tasks helps to explain how the jurisdictional problems came about in the first place.

The services acquired by DMF in 1904 pertained to aids to navigation, including the charting of river and coastal waters. Hydrographic surveys, however, were also essential for marine engineering works such as the construction of harbours and canals. These surveys continued to be undertaken by DPW and DRC. Still other forms of hydrographic data were required for planning irrigation works of western farming and for identifying water-power potentials for hydroelectric generation. The Department of the Interior (DOI) and DPW were the two federal departments concerned with these non-navigation hydrographic activities.

Marine and Fisheries, established at Confederation, provided aids to navigation and managed public harbours.[1] Survey work focused on the need to supply charts and related aids to navigation. For the remainder of the century, the British Admiralty provided much of the navigation hydrography that enabled the DMF to fulfill its mandate. This responsibility declined after the creation of the Canadian Georgian Bay Survey in 1883, but the Admiralty continued to conduct surveys until 1910.

The Department of Public Works, also created at Confederation, had a far broader mandate to construct marine works, such as canals and harbours, and to improve navigable waterways. The department also operated the canal system. Both DPW and DMF acquired their facilities and hydrographic skills from the various former colonial public works departments.[2] In 1879 the DRC was hived off from DPW in recognition of the growing importance of railways and, to a lesser extent, the increasing busyness of inland waterways.[3] Broadly speaking, DRC had jurisdiction over artificial waterways, while DPW retained control over natural waterways.

In addition to navigation, hydrographic surveying was a component of assessing water powers, irrigation potentials, and flood control. Many of these functions were associated with the expansion of settlement in western Canada. The Department of the Interior (DOI) was created in 1873 to manage the affairs of the recently purchased North-West Territories.[4]

Despite this seemingly straightforward functional division of responsibility, considerable jurisdictional overlap evolved. Hydrography is the study of the physical character of the beds of water bodies, currents, water volumes, and other data associated with water in a "natural state."[5] The boundary between hydrographic surveys for engineering and aids to navigation quickly became a grey zone. Where was the demarcation between a survey conducted exclusively for engineering purposes and a survey to aid public navigation? Over time, DMF acquired more and more of the marine engineering functions of DPW. Conversely, DPW and DRC made some of their surveys available to mariners. Ten years after Confederation, the government felt it was necessary to pass clarifying legislation regarding the construction and operational responsibilities of the two departments.[6] However, the act resolved little.

In 1904 several issues conspired to bring navigation hydrography under one agency. Specifically, the grounding of the steamer *Sicilian* on 24 November 1902 in the St Lawrence River showed that too many federal departments were involved in charting waterways. The 1892 legislation reorganizing the DMF had reiterated that all tidal observations and hydrographic surveys were under the direction of the DMF chief engineer. Yet this did not stop DPW from creating a St Lawrence hydrographic survey in 1896. More significantly, in the early 1900s the British Admiralty wanted the Canadian government to assume all responsibility for its hydrographic surveys. In March 1904 the hydrographic survey work of DPW, DRC, and DMF were combined into a Canadian hydrographic service within DMF. In May the Admiralty made a formal request that Canada assume the former British obligations.[7]

The 1904 transfer did not completely end the dispersion of hydrographic services in aid of navigation. DPW continued to install and monitor automatic water-level gauges on the St Lawrence River and the Great Lakes. It was not until 1912 that these were transferred to the Hydrographic Survey.[8] Similarly, as long as new engineering projects were being built, detailed surveys were essential and not part of the purview of the new Hydrographic Survey of Canada. DPW was still conducting river surveys at least up to the First World War.

Although the departments undertook hydrographic surveys for different end results, the methods were the same. The main difference between navigation and engineering surveys was in scale. Navigation charts often provided the starting point for marine engineering works. Engineering design and cost estimates depended on knowledge of hydrographic features such as the character and contours of the bed of the water body and the tides and currents. The small scale and limited coverage of navigable waters dictated that there was a continuous domestic demand for soundings, current and tide studies, and other hydrographic work at large scales of very specific geographic areas such as a harbour or river channel. Rear Admiral Wharton, hydrographer of the Royal Navy in Britain, described three levels of hydrographic surveys, namely, preliminary or sketch surveys, surveys for the ordinary purposes of navigation, and detailed surveys.[9]

Marine engineering was the biggest activity of the DPW budget. In 1903, just prior to the transfer of hydrographic services to DMF, the department spent the following:

| | |
|---|---|
| Harbour and river work | $2,376,000 |
| Dredging | 982,000 |
| Timber slides and booms | 136,000 |
| All other (buildings, bridges, roads, telegraph etc.) | 2,337,000 |

Dredging was the single most frequent maintenance and construction task undertaken for harbours and rivers. DPW began systematic dredging programs in 1873, as the country needed deeper harbours and improved channels.[10] These waterways had to be maintained in a safe condition and enlarged as deeper-draft vessels were introduced.

During its first fifteen years of existence, DPW assisted in improving 307 ports, harbours, bays, and rivers across Canada. Improvements ran the gamut from pier repairs to vast breakwater projects, and almost all involved some amount of dredging. This work ranged in scale from $200 spent in 1881 to "deepen" the entrance channel at Little Harbour, Nova Scotia, to the immense, ongoing task of enlarging and maintaining the St Lawrence River from Quebec to Montreal. By 1882 the government had invested $2.9 million in this project.[11]

Calculating the cost and time of a dredging operation required a detailed knowledge of the topography and character of the bed of a water body. Soundings to determine bottom contours were typically made at 7.6 metre (25 foot) intervals along lines spaced on 15 to 30 metre (50–100 foot) centres, depending on the survey purpose and bottom irregularities. For example, soundings undertaken by DRC in Halifax Harbour in 1912 were done at 7.6 metre intervals on lines placed 30 metres apart, but the department surveyed a rock shoal in the St Lawrence at Montreal in 1880 at 3 metre (10 foot) intervals.[12] Soundings were normally made with lead-lines taken from a boat. In the winter, surveyors could take soundings from the ice on rivers and harbours, dropping the lead-line through holes cut in the ice at appropriate intervals.[13] Soundings were typically taken before and after a dredging operation to calculate the actual quantity of dredged material. This was known as "place measurement." The alternative was to measure the volume of dredged material dumped into scows, a method known appropriately as "scow measurement." A minimum and a maximum dredge depth were usually specified. The maximum depth, often 60 centimetres (2 feet) below the specified bottom, discouraged contractors from excavating, and being paid for, more material than necessary.[14] Dredged material had to be removed from the work site, and a separate survey was needed to locate a dumping ground.

Virtually all dredging involved natural material. One man-made problem was sawmill waste, primarily sawdust, and it reached such a state that legislation was passed in 1873 to prohibit dumping of mill wastes. The seriousness of sawdust depended largely on hydraulic conditions. For example, although saw milling was a major industry on the Saint-Maurice River in Quebec, the fast river flow kept the channel clear. On the other hand, confused currents at the Chaudière Falls at Ottawa-Hull or lack of adequate current, such as at Belleville harbour in Ontario and Chicoutimi harbour in Quebec, caused problems that required regular dredging.[15] The variables involved in trying to determine accurately the character of the invisible bed of a water body meant there was always room for dispute between the government and the contractor. These disagreements were usually resolved within the normal process of arbitration or the courts, but some had immense ramifications for the DPW. The most serious controversy involved dredging in the harbour of Quebec City between 1882 and 1888. Collusion between the contractor, Larkin, Connelly and Company, and some politicians, particularly Thomas McGreevy, produced huge profits on dredging contracts. Patronage was a tolerated way of doing business with the government, but the activities surrounding this work far exceeded any of the unwritten limits. In 1891 Parliament laid formal charges against McGreevy; he was found guilty and expelled from Parliament. Hector Langevin, the minister of Public Works, and Henry Perley, the DPW chief engineer, were forced to resign.[16]

The scale of the St Lawrence River dredging operation below Montreal holds a unique place in the development of Canada's marine infrastructure. At the end of the nineteenth century it was considered the most important dredging work in North America.[17] Prior to any improvements, the ruling depth of the St Lawrence was found in Lac Saint-Pierre, a widening of the river between Sorel and Trois-Rivières. The lake contained a 10-kilometre stretch with a natural depth of 5.5 metres (11.5 feet), and only vessels of 400 tons or less could cross Lac Saint-Pierre and hence reach Montreal from Quebec or the Atlantic Ocean.[18] As early as 1826, Montreal merchants had petitioned the Lower Canada legislature to deepen the river, but it was not until 1844 that dredging began. This proved to be a failure, possibly because of difficult soil conditions or river currents. Dredging was abandoned in 1847 after a considerable sum of money had been expended.[19] Montreal merchants found this situation untenable, and in 1850 the Montreal Harbour Commissioners took over the work. By the end of 1851 the commissioners had achieved a channel 4 metres (16 feet) deep. As ocean shipping was continually increasing in size, the commissioners immediately proceeded to increase the channel to 6.1 metres (20 feet). This was completed in 1865. In 1878 the channel had reached a depth of 6.7 metres (22 feet); by 1888 it was 8.4 metres (27.5 feet). In 1888 DPW assumed this dredging responsibility from the Harbour Commissioners, and the work has remained a federal public work ever since. In 1904 management and control of dredging the ship channel was transferred from DPW to DMF.[20]

The British North America Act gave the federal government responsibility for "Beacons, Buoys [and] Lighthouses" and also for "Navigation and Shipping." The concept of a navigable waterway has changed both over time and geographically. In areas of rail competition, vessels had to become larger and more specialized if the waterway was to remain commercially viable, whereas on frontier rivers relatively simple vessels could operate profitably. An immense number of waterways were considered navigable in 1900, but commercial activity on each route was highly variable.[21] Waterway improvements evolved in two directions: those that were intended to provide permanent transport infrastructure and those that were stop-gap measures until better transport was available.

The Saint John River in New Brunswick is an example of a regional waterway that, with improvements, was an important navigable route for many years. It was navigable in three sections separated by short, non-navigable sections. The most important was a 130-kilometre passage from Saint John to Fredericton. River steamers to a maximum size of about 190 feet (58 metres) in length used it until the last vestiges of commercial traffic shifted to roads in the 1930s. The second section, from Fredericton to Grand Falls (225 kilometres), was navigable to smaller steamers, but the passage was greatly affected by extremes of high and low water. From Grand Falls to the mouth of the St Francis River (120 kilometres) the river was very shallow and used only by towboats. Both the DPW and the provincial government made considerable improvements to all sections of the river.[22]

The Saskatchewan River, by contrast, was cheaply and quickly improved in the 1880s to provide basic transport across the prairies until railways could be constructed. In 1857 Henry Youle Hind, while on his exploratory survey of western Canada, commented on the navigability of the Saskatchewan River: "If the channel were cleared of boulders and improved, it might be ascended by a powerful steamer."[23] Little happened until after the dominion government purchased the Hudson's Bay Company lands in 1869 and created the province of Manitoba the following year. The HBC placed steamers on the Saskatchewan

River in 1874.[24] In 1880 the company indicated that it would continue a steamer service on the Saskatchewan as long as DPW would improve the river. Public Works was not anxious to undertake the work and instead, in 1882, provided the HBC with the necessary funds. The improvements, consisting mainly of boulder removal, were finished in 1884. Railway competition ended the steamboat service five years later.[25]

Alternatives to dredging included raising water levels to flood obstructions and constructing overland portages. The Yamaska River in Quebec became navigable from the St Lawrence for 32 kilometres when a dam was erected to flood a shallow section. A lock carried vessels around the dam. Similarly, between 1887 and 1895, DPW built a lock and dam to flood a small rapid on the Lièvre River in Quebec.[26] Tramways to portage freight around rapids were used on several river systems, such as the Athabaska and Yukon Rivers. One of the longest lasting operations ran from 1877 to 1922 at Grand Rapids, where the Saskatchewan River empties into Lake Winnipeg.[27]

Water routes frequently combined rivers with canals and locks. They required both a hydrographic survey to determine channel depth and character and hydraulic studies to devise a smooth flow of water through the built structure. Whereas the St Lawrence River improvements below Montreal were largely ones of dredging, making the river navigable above Montreal required canals. In 1848 the river finally became navigable to steamers by means of seven canals at Lachine, Beauharnois, and Cornwall and four short canals between Prescott and Morrisburg, known collectively as the Williamsburg Canals. The Galops Rapids (bypassed by some of the Williamsburg Canals) near Prescott, Ontario, contained a challenging combination of hydraulic and hydrographic problems. Originally, two canals, the Galops and the Iroquois, were built on this section of river. But because of an engineering error, the water level was too low in the Iroquois locks, and in 1851 the two canals were connected by a dike and converted into a single canal. Beginning in 1876, DPW, and later DRC, undertook extensive channel improvements to deepen the approach to the upstream end of the canal. The enlargement was supposed to have been completed in 1888, but work stopped prior to that because of an apparent permanent lowering of the St Lawrence River in that area. It was not completed until after a river survey was made in 1897.[28]

Dredging approach channels to locks was a necessary part of canal maintenance. DRC maintained dredges, tugs, and scows on the Rideau and Trent Canals. Sometimes waterways had to be dredged when locks were built to handle bigger vessels. For example, a channel providing 17 feet (5.2 metres) of draft was excavated through Lac Saint-Louis on the St Lawrence River between 1895 and 1899.[29] Dredges were also used to enlarge channels between locks when canals were rebuilt, as in the case of the Lachine Canal in the 1890s.

Haphazard construction of the Trent-Severn Waterway created what must have been the most jurisdictionally complicated navigation system in Canada. The watershed was used for both navigation and lumbering, and improvements consisted of building individual locks linking short sections of navigable waterway and erecting timber slides. The dominion and Ontario governments and private lumber companies also maintained a complex system of water regulation structures. By 1873 most of the reservoir dams were under the control of the Ontario Department of Public Works. In 1880 about 44,600 hectares (110,000 acres) of reservoirs were being maintained in the watershed. In 1885–86 DRC undertook a watershed survey to try to ameliorate navigation problems caused by competition with timber companies for water. By 1900 over fifty dams stored 28,400 hectares (70,000 acres) of water.[30] Water regulation was not satisfactorily resolved before 1914.

After 1900 Canada entered a long period of economic prosperity. Business and political optimism were so great that there was a fear that the existing commercial waterways (and railways) would not be able to handle all the new traffic. One proposed method of increasing carrying capacity was to revive the long-proposed Georgian Bay ship canal. As early as 1837 the Upper Canada legislature had authorized a survey of a canal linking Montreal by way of the French and Ottawa Rivers to Georgian Bay and hence providing a shortcut for shipping between the upper lakes and the St Lawrence. A second survey was conducted in 1857. Nothing came of these grand schemes, but more modest surveys of short sections of the route were undertaken at the end of the century. In 1896 DPW undertook a survey with the limited objectives of connecting Georgian Bay by way of the French River to Lake Nipissing to improve the grain-handling capacity of the Great Lakes system. Later, in 1899–1900, DRC conducted a hydrographic survey to evaluate the feasibility of enlarging the navigable section of the Ottawa River from the St Lawrence to Ottawa-Hull.[31]

The concept of the Georgian Bay ship canal in its entirety re-emerged when DPW was authorized in 1904 to undertake the most extensive and expensive hydrographic survey of the route. The report in 1908 indicated that while such a canal was technically feasible, it would cost $100 million and take ten years to build. The Liberal government was committed to construction of transcontinental railways. The canal was far too expensive, and the idea finally died.[32]

The federal responsibility for navigation and waterways can be seen to cover two distinct types of water bodies. River and canal transportation routes, being in confined channels, were unlikely to require lighthouses and harbours of refuge or to be maintained during the winter season. Coastal navigation, on the other hand, occurs on open water, with all the uncertainties of weather and geography. The navigable coastal waters may be divided into four regions: Atlantic, Pacific, Hudson Bay, and the Great Lakes.

On the Great Lakes the passage between Lake Huron and Georgian Bay was an example of the navigation challenges faced in coastal waters. A 29-kilometre barrier of rocks, reefs, ledges, shoals, inlets, and islands extends from the tip of the Bruce Peninsula to Manitoulin Island, separating Lake Huron from Georgian Bay. Only two shipping channels passed through this maze of shoals.[33] As lake traffic increased through the ports of Collingwood and Owen Sound, this obstacle became serious. The Main Channel, as its name implies, was the most important entrance into Georgian Bay. The Board of Works completed the Cove Island lighthouse in 1859 to mark the Lake Huron entrance to this channel.

In 1870 DMF assumed responsibility from DPW for constructing lighthouses that cost less than $10,000. Since the great majority of coastal lights were built for less than $4,000, this arrangement was presumably made to simplify administration by having the department that requested, and subsequently operated, the structure also responsible for its construction.[34]

While lighthouses provided navigation guidance to shipping, "harbours-of-refuge" from storms were maintained for emergencies. Prior to the advent of a predictable meteorological service, captains had no long-range warning of storms. Being caught in a bad storm close to shore was especially risky for sailing vessels, as they lacked the manœuvrability of steam-powered ships.[35] Commercial harbours were also harbours-of-refuge if they could be entered safely during a storm. Along Lakes Erie and Huron the character of the shoreline limited the number of safe harbours. The absence of adequate harbours-of-refuge was so critical that one of the earliest DPW hydrographic surveys was undertaken in 1869 to assess possible harbours-of-refuge along the shores of these two Great Lakes.[36] Even when harbours existed, they could not always be improved for protection from a storm. The base of Long Point on Lake Erie was an especially dangerous place for a sailing vessel in bad weather because of the prevailing winds and lake currents. Although Port Burwell was ideally located at the base of the point, it could not be entered in a storm and hence was never identified as a harbour-of-refuge. Along unpopulated shorelines, such as Georgian Bay in the 1880s, efforts were made to find natural harbours-of-refuge. One of the purposes of the Georgian Bay Survey in 1883 was to locate suitable anchorages during storms.[37]

Whereas river navigation was inevitably closed during the winter, considerable political and economic incentive existed to extend the winter navigation season on coastal waters. Just as the Maritime provinces had earlier been linked to central Canada with a railway, so the dominion government promised Prince Edward Island a year-round ferry link with the mainland when it entered Confederation in 1873. The Northumberland Strait, between the island and the mainland, is only 14 kilometres at its narrowest. In winter the strait froze but, because of tides and currents, not as a solid sheet. Keeping a shipping channel open through the ice therefore appeared feasible. Between 1876 and 1909 DMF placed a series of icebreakers on this route. However, it was not until DRC introduced a large train ferry in 1918 that a reliable service could be guaranteed.[38]

Like so much else about the river, ice on the St Lawrence created unique engineering conditions for the DPW. Much of the departmental work focused on reducing flooding caused by ice dams. In 1898 the well-known engineer Thomas Keefer described the history of floods on the river and examined the mechanisms by which different types of ice were created and the relationship between ice types and the shape of the river channel and its currents. The most intriguing part of his report was the hint that winter navigation of the St Lawrence might be possible with suitable icebreakers. Although flooding caused by ice jams was largely eliminated by 1900, winter navigation on the St Lawrence did not begin until the 1980s.[39] The last third of the 1800s marked the pioneer period of ice breaking in Canada. This priority was in response to several issues: railways needed year-around marine services, flooding on the St Lawrence had to be reduced, and the Confederation promise to Prince Edward Island had to be honoured.

Submarine telegraph laying was one other service in coastal waterways that required hydrographic surveys. A dominion government telegraph system comprising land lines and submarine cables was proposed as early as 1876 but not actually established until 1880. The telegraph service was to facilitate the gathering of information for ocean shipping, the fishing industry, and meteorological studies primarily in the Gulf of St Lawrence. In 1880 the dominion government purchased privately owned lines in British Columbia to establish a similar service on the Pacific coast. Most of the two systems were land lines, but by 1882 DPW was responsible for 243 kilometres of submarine telegraph cable.[40] The Dominion Telegraph and Signal Service continued to build routes that were unprofitable for private companies.

Public Works and Marine and Fisheries were the two dominion departments specifically charged with harbour development. In practice, however, few if any Canadian ports were under the direct authority of the federal government. Instead, port ownership and operation were under the authority of a variety of semi-public and semi-autonomous harbour commissioners and harbour masters.[41] Harbour commissioners maintained and improved harbours through dredging and adding new infrastructure. The first harbour commission seems to have been created at Montreal in 1830; the one at Quebec was organized shortly after. By the early twentieth century, harbour commissioners controlled fourteen of the most important ports in Canada. In theory, they were independent bodies, but in fact, DPW often acted as overseer of their projects. This control existed because harbour commission projects relied on funds or loan guarantees voted by Parliament. Rather than lose complete control of federal funds, DPW exerted a secondary control on expenditure, while the commissioners remained responsible for the actual planning and direction of projects. The confused relationships often made it difficult to tell exactly where the crucial decisions regarding the work were made. Henry Perley, for example, worked for the Quebec Harbour Commissioners while he was also chief engineer of Public Works.[42]

The most significant period of port expansion occurred during the twenty prosperous years preceding the First World War. The country needed more transport infrastructure to get goods from inland to the coast and from there through ocean ports to Britain.[43] Between 1900 and 1910 the federal government annually increased expenditures on harbour facilities specifically dedicated to the grain trade.

With completion of the Intercolonial Railway between Halifax and Lévis (opposite Quebec City) in 1876, the dominion government became more involved in the improvement of the Port of Halifax. The harbour was ice-free throughout the year and became increasingly important as a winter port when the harbours of Montreal and Quebec were frozen. In 1881 DRC acquired lands for a "deep water terminal" closer to the commercial centre of Halifax than the original Intercolonial terminal.[44] Over time, the department expanded the harbour facilities at this site.

The ports of Halifax, Saint John, Quebec City, and Montreal and several Great Lakes ports were all slated for massive expansions. Of all of these plans, only the Halifax "ocean terminals" development was completed substantially as contemplated. The DRC hydrographic survey of the harbour in 1912 marked the commencement of this construction.[45] Saint John was also an ice-free port but not as well-endowed with a natural harbour as Halifax. In 1874–75 DPW made a thorough hydrographic survey of the harbour. Subsequent work included the construction of a large breakwater and dredging. The department continued to make large improvements until well after the First World War.[46] DPW constructed some unusual harbour structures to deal with local conditions. Between 1876 and 1888 it built a wet dock – the only such structure in North America – at Quebec City. The dock was built to address concerns of tides, ice, and river currents.[47]

The Great Lakes harbours of Port Arthur and Fort William (collectively the Lakehead and today Thunder Bay) were the pivotal point of an intricate rail-water transport system for moving prairie grain to the ocean export ports of Montreal, Saint John, and Halifax. Just prior to the war, the two harbours received additional protection with the construction of long breakwaters, and the Kaministiquia River was dredged at Fort William to increase berthing space. By 1922 the Lakehead had the largest grain elevator storage capacity in the world.[48] Most Great Lakes ports, such as Collingwood and Kingston, were much smaller. Toronto's harbour was the largest general harbour on the Canadian lakes. DPW conducted a hydrographic survey in 1881 to determine how to improve the harbour for shipping. Dredging began in 1882, and by 1887 most of the work had been completed.[49]

In the 1860s Victoria, on Vancouver Island, and New Westminster, at the mouth of the Fraser River, were the two important Pacific coast ports. Neither was especially good in its natural state. Victoria Harbour was very shallow, and DPW undertook a dredging program during the 1870s. Similarly, dredging of the Fraser River in 1879 shortened the distance from New Westminster to the mouth of Fraser River by three kilometres for boats drawing up to three metres.[50] Immediately upon British Columbia entering Confederation in 1871, DPW had commenced route surveys for the Canadian Pacific Railway, but the Canadian government waited until 1876 to ask the British Admiralty to comment on the selection of a Pacific coast port. By then, railway surveys had identified three possible terminals, each reflecting a compromise of hydrographic, political, commercial, and engineering constraints.

The Admiralty report was completed by November 1877. The most northerly location was Port Simpson, about thirty kilometres north of the mouth of the Skeena River and near the future boundary with the United States. Port Simpson was well known as an anchorage and seemed to have all the requirements for a terminal harbour. A port at the mouth of the Skeena River would have saved many kilometres of expensive track construction. The only decent anchorage was located at nearby Cardena Bay, but it was technically impractical to construct a line to this location. The Admiralty report therefore considered Port Simpson and the mouth of the Skeena as unsuitable harbours.[51] The second route was a line down Bute Inlet and then an island-hop with bridges to Vancouver Island and a terminal at Victoria or Esquimalt. The dominion government had promised British Columbia an island terminal as a consideration for joining Canada. Prior to the Admiralty survey, the railway engineering staff knew that the bridges would have to be exceedingly long and built over deep channels with strong currents. Marcus Smith, who was in charge of the British Columbia railway surveys, championed this alignment even after the Admiralty and Sandford Fleming, engineer-in-chief of the railway survey, had dismissed the route.[52] The third possible line was down the Fraser River to Port Moody at Burrard Inlet. This site was the preferred Admiralty port. Fleming also favoured the Burrard Inlet terminal and, to a lesser extent, the Port Simpson harbour. In 1879 the dominion government selected the Burrard Inlet terminal, thus ensuring the future development of Vancouver.[53]

In 1906 Admiralty hydrographers resurveyed the area around Port Simpson and the mouth of the Skeena River to locate a terminal for a new transcontinental line, the Grand Trunk Pacific Railway. Again, the survey determined that no practical anchorage existed on the Skeena. However, according to one tale, a wonderful harbour known as "Tuck's Inlet," near the mouth of the Skeena, intrigued the railway surveyors. According to Admiralty charts, a submerged shoal blocked the entrance, but the railway survey was unable to find any evidence of a reef. Subsequent research showed that the original charts had inadvertently marked this reef, which was in fact located in an adjacent bay. "Tuck's Inlet" became the railway port of Prince Rupert.[54]

The Hudson's Bay Company had maintained ports on Hudson Bay since the company was founded in 1670. With prairie settlement booming at the end of the nineteenth century, there was interest in developing a Hudson Bay port as an alternative ocean outlet to the St Lawrence. DMF sent exploring parties into Hudson

Bay in 1884, 1885, and 1886. During 1886, surveys of the Churchill and Nelson harbours were undertaken. These surveys concluded that Churchill was far superior to the mouth of the Nelson River. The mouth of the Nelson was shallow, and sediment brought down by the river would quickly silt the harbour.[55]

In 1908 DRC conducted a survey for a proposed Hudson Bay railway. As expected, the terminal was to be Churchill, but in the fall of 1909 a separate survey was conducted to Nelson. Surprisingly, the DRC chief engineer recommended Port Nelson as the ocean port. This choice seems to have been based on the quality of the land and the shorter route rather than the suitability of the harbour. The department's selection of Port Nelson compelled DPW, on short notice, to conduct a reconnaissance survey of the Nelson River to determine if it could be made navigable between Lake Winnipeg and Hudson Bay.[56] In 1913 DRC officially approved Port Nelson,[57] but fortunately, construction stopped because of the war. Before work resumed, new hydrographic information highlighted the total unsuitability of Port Nelson; the railway was rerouted to Churchill.

Survey work is normally associated with the publishing of maps or charts. Because of their diverse authorities over land and water issues, DMF, DPW, DRC, and DOI were the principal map- and chart-producing departments of the dominion government.[58] With regard to hydrographic work, it appears that because of the very specialized nature of projects undertaken by DPW and DRC, most of their data remained in manuscript form. Manuscript maps were produced at a variety of scales. For example, a DPW survey of a portion of the St Lawrence River undertaken in 1875–76 was plotted at 1:4800 (400′:1″). In 1903 hydrographic charts of the St Lawrence River channel were being plotted at 1:6000 (500′:1″). Relatively little of the material appeared as publicly available navigation charts.

There were exceptions. In 1884 DRC commenced a hydrographic survey to provide estimates for improving the Richelieu River navigation. When the work was completed at the end of 1887, publication of the survey was considered, but it does not appear to have been issued.[59] DRC did publish a navigation chart for Lac Saint-Louis after completing a dredging project in 1899.[60] Overall, the extent to which hydrographic data generated by DPW and DRC were used by other government agencies or the private sector has not been examined in this chapter.

As part of its ongoing dredging operation on the St Lawrence River, DPW established a hydrographic survey unit in 1896 to chart the St Lawrence between Montreal and Quebec City. The purpose of the survey was to map the existing 8.4 metre (27.5 foot) channel depth and estimate costs of increasing it to 9.1 metres (30 feet) and widening the channel beyond 91 metres (300 feet). By the turn of the century, the survey had begun drafting navigation charts. The first chart was ready for printing when the DPW hydrographic unit was transferred to DMF in 1904. It was issued in 1905 as *Montreal Harbour, Longue Point to Varrennes*.[61]

In addition to aiding river and coastal navigation, hydrographic studies were undertaken for a variety of non-navigation functions. In the twentieth century many of these functions would be placed under provincial jurisdiction. Identification of potential water powers was a task that became increasingly important with the development of hydroelectric power after the 1890s. Determining irrigation potential for farming arid western lands was another area of work. Identifying and rectifying flood potential became increasingly important as humans altered the natural water regimes.

In the 1800s, water power was used primarily to generate mechanical power to drive mills to grind grain, cut lumber, produce textiles, and operate other small manufacturing enterprises. Mills were found on both navigable and non-navigable rivers. Those on non-navigable waterways were subject to riparian rights of landowners but not to specific federal jurisdiction. On navigable waterways, mills were found adjacent to dams used to create deep-water navigation. Extra water, not needed to operate the locks, was available by lease from the federal government. Significant amounts of mechanical power were developed on the Welland, Rideau, Trent, and St Lawrence River canals. By virtue of dominion control over navigable waterways, DPW and later DRC were involved in developing and leasing water powers. However, the needs of water power and navigation were not complementary. When canal water levels were low, water was cut off to the mills to ensure that shipping could continue to move.[62]

The water powers of the Lachine Canal at Montreal and the Chaudière Falls at Ottawa-Hull became the two most developed sites. DPW began leasing water power on the Lachine after it rebuilt and enlarged the canal in 1848. The Lachine power was especially well located as it provided cheap power in the centre of Montreal's industrial district at a time when steam power was expensive.[63] DPW never received full value for its leases since payment was based on what should have been used and not on what actually was consumed. Without adequate regulation, mill owners drew off as much water as they could, often to the detriment of shipping. The Chaudière Falls was the only site to approach the Lachine Canal in terms of scale of power generated. In 1851 the Board of Works carried out a hydraulic survey at the falls. Dams were built in 1854, and water lots sold to sawmilling companies.[64] By 1867 water power leased in this small area was about the same as all the water power on the Lachine Canal.

The demand for mechanical water power paled beside the huge twentieth-century requirements for hydroelectric power. Hydrographic data collected over several years became essential to ensure a reliable and predictable source of water. With few exceptions, water powers on navigable waterways were quite small and contributed a negligible amount of power. The Dominion Power and Transmission Company's 1898 DeCew hydroelectric plant was one of the exceptions since it drew its water from the Welland Canal and used the Niagara Escarpment to create a 270-foot head.[65]

With the increasing size and economic importance of cities such as Montreal and Winnipeg, the effects of floods were a serious and obvious problem. But flood control was not just an urban problem. With the advancement of settlement across Canada, large tracts of land suitable for settlement were affected. For example, Lac Saint-Jean, Quebec, had a large drainage basin but a modest discharge channel into the Saguenay River. Thus spring runoff could cause significant flooding of its shoreline. The lake could rise rapidly by 4.6–6.1 metres (15–20 feet) and occasionally to 9.1–10.7 metres (30–35 feet). In 1880 the bishop of Chicoutimi asked DPW to enlarge the Grande Décharge exit of the lake. This would increase speed of discharge and reduce flooding to settlements around the lake. DPW undertook the work because it was already maintaining timber slides on this section of the Saguenay River.[66]

The absence of hydrographical information at Lac Saint-Jean illustrated the problems facing engineers in the new country as they tried to provide solutions. "As there is no correct chart of Lake St. John as yet, it is very desirable that the lake should be surveyed and sounded, and that the supply of water from its tributaries should be ascertained, together with the outflow through the Grande and Petite Décharge. These should be measured, sounded and leveled, with accuracy along the portions which obstruct the discharge of water during the spring floods."[67]

The Department of the Interior held a special place in the hydrographic surveying. In 1880 it made a request to DPW to study flooding of Lake Manitoba. The lake's water level had been increasing over time and threatened the viability of farms along the shore. The initial DPW survey noted that there was insufficient hydrological data to understand the lake's water regime but that dredging the Fairford River, at the exit from the lake, might lower the water level. The following year new data determined that the river bottom was rock and dredging was not economically feasible. Some limited, and unsuccessful, dredging was undertaken at the turn of the century.[68] Regulation of Lake Manitoba's water level remains a problem today, but it is now a provincial, rather than federal, responsibility.

Irrigation surveys became essential to settlement in dry regions of western lands under DOI jurisdiction. By the early 1890s the department had identified the necessity of undertaking a topographic survey in preparation for large-scale irrigation projects, especially in what later became Alberta. Passage of the *North-West Irrigation Act in* 1894 provided the authority to proceed with a survey to estimate the volume of water available for irrigation.[69] Revisions to the *Irrigation Act* in 1898 gave DOI regulatory powers over irrigation. Presumably in recognition of the department's lack of engineering expertise, the act made the DPW chief engineer also the chief engineer for DOI irrigation projects. The work of stream measurements became so demanding that in 1909 DOI established a Canadian Hydrographic Survey[70] (it later became the Hydrometric Survey).

The demands for federal hydrographic services changed after the First World War. Data collection for navigation charts and other aids to navigation remained important but was now clearly within the realm of the Hydrographic Survey of Canada. However, the 1904 order-in-council creating this body did not end the hydrographic work of DPW and DRC, and they continued to undertake marine engineering projects. Nevertheless, as the number of navigable waterways declined in the face of road and rail competition, the hydrographic expertise of DPW and DRC to maintain them was greatly diminished. Finally, vestigial federal jurisdiction in several areas of hydrographic activity, including irrigation, flooding, and water powers, was transferred to the provinces in the early part of the century. The changing hydrographic activities in Canada had evolved with the growing sophistication of the country's political and transportation systems.

# 6

# *HMS* Challenger's *Surveys in Labrador, 1932–1934*

G.S. RITCHIE

Until 1949 Labrador, as a part of Newfoundland, was a British colony.[1] Providing hydrographic surveys to improve navigation along the coastline was therefore the responsibility of the British government. In the late 1920s the Colonial Office had been pressing for surveys to be made so that an inshore route northwards from Newfoundland could be charted. Such a route would enable vessels to travel to the coastal settlements four or five weeks earlier in the summer, before the pack ice melted sufficiently to allow navigation in the open sea clear of the labyrinth of rocky islands. At the same time surveys were to be made to permit safe passage from the main coastal settlements through the clusters of islands, reefs, and pinnacle rocks to the open sea.

In 1930 the Hydrographer of the Navy arranged to make such surveys, and for the work he selected the newly built survey ship *Challenger.* She was a single-screw steamship of about 1,200 tons with a draught of approximately 12 feet and just over 200 feet long; she was broad in the beam with a high forecastle, and her stern had an overhanging counter. She was a lively but dry-sea boat, even in the heaviest weather. Commander Guy Wyatt, a senior hydrographic specialist with many years of experience, was in command of *Challenger* for the Labrador surveys, while

HMS *Challenger* off the coast of Labrador, 1932

Lieutenant-Commander E.H.B. "Buck" Baker, as first lieutenant, headed a team of surveying assistants.

The ship was fitted with the recently developed British Admiralty pattern echo-sounding apparatus made by Henry Hughes and Son Ltd of London. A sonic signal induced by the striking of an electronically controlled hammer three times a second on a metal plate within the transmitter secured to the ship's hull provided the outgoing acoustic signals. The returning echoes from the seabed were transferred from the receiver on the ship's hull to the timing switch, which in turn controlled the measuring wheel graduated to reveal the depth of water beneath the keel. The operator, wearing earphones and seated beside the apparatus on the bridge, rotated the wheel to and fro until he heard the loudest signal, when he called out the depth as shown by the pointer. The ship carried two motor sounding launches, from which depth soundings were taken by two leadsmen, one each side of the boat, using lead and line; no echo sounders were yet available for these boats.

On Thursday, 7 July 1932, the ship, loaded with stores and fresh provisions and with the mails for northern Labrador, sailed from St John's, Newfoundland, for Hopedale. By nightfall the following day, the ship was off Belle Isle, icebergs and growlers having been sighted during the day, which had been fine and clear with a cloudless sky. As the ship sailed northwards off the

The British Admiralty pattern echo sounder

coast, an increasing number of bergs were seen, between twenty-five and fifty being in view at any one time. Captain Clarke, the embarked pilot, said that this number was usual along this part of the coast at this time of year. He had been navigating these waters as a fisherman and as a master of sealers and icebreakers and, finally, of the Newfoundland government steamers for over forty years and was now sixty years of age. Bergs were becoming so numerous that navigation after dark was hazardous, and the ship anchored for the night of Saturday, 9 July, in Webeck Harbour, passing into the anchorage between two icebergs grounded in the entrance.

At dawn the ship weighed and stood over to the south side of Ragged Islands, passing between them and a small double island, thence past Black Bear Island and northeastward of

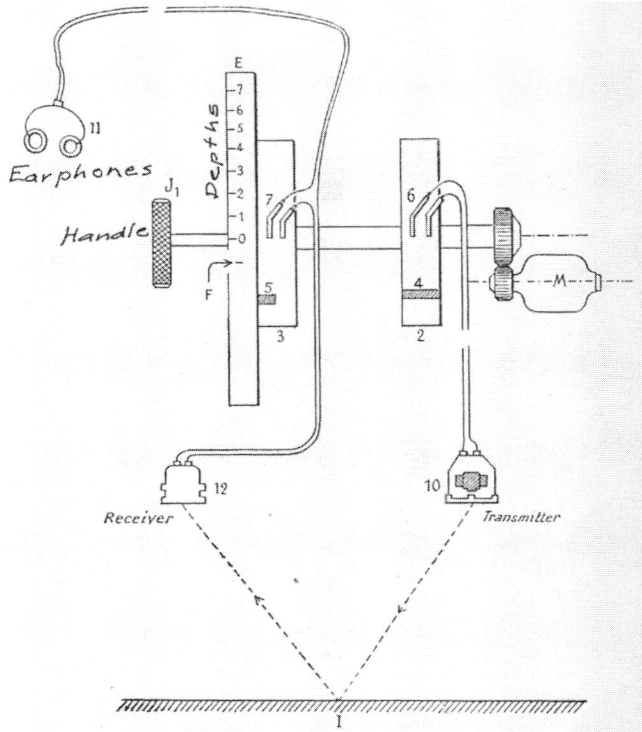


Diagram of the Admiralty pattern echo sounder

Mortimer Shoal, when the wind shifted to the northeast and fog came into the land. This being long before the days of radar, the ship had to look for an anchorage and just found one in a bight at the south end of Kidlialuit. By noon the fog had lifted again, and weighing anchor, the ship passed between the Iron Bound Islands and then across between Makkovik Island and Ulgoklialuit Island. She then stood over to Cape Makkovik and between the Turnaviks, south of Striped Island and round Tikkerasuk, and across to Kayaksuatilik, where the passage between the islands and rocks is exceedingly narrow though quite deep. The existing chart was almost unrecognizable, but from here across to Hopedale was a nearly straight run with pack ice visible to seaward. By 7 p.m. the ship was at anchor in Hopedale, and the captain landed to call on the local missionary employed by the Grenfell Mission.

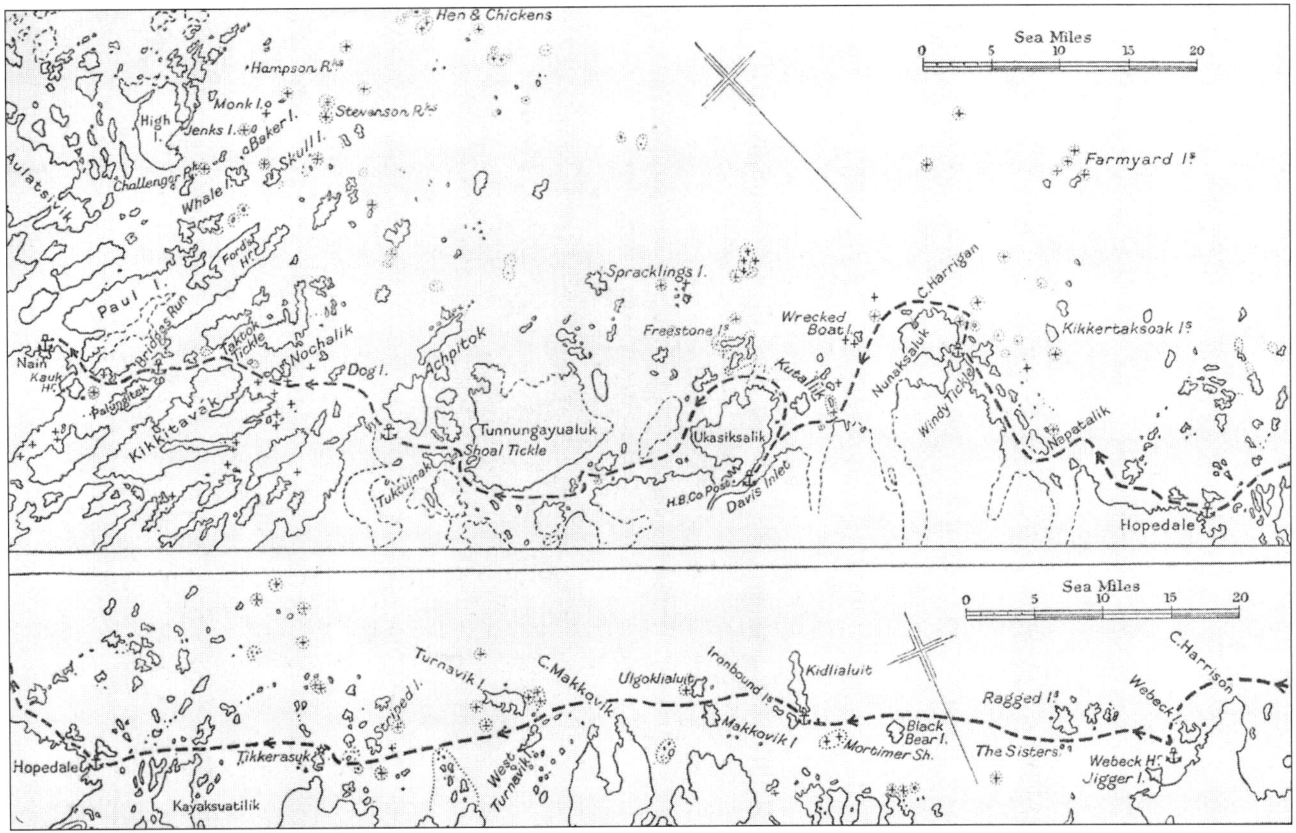


Track chart of HMS *Challenger*

Breaking through pan ice

At noon on Monday the ship again proceeded northwards for Nain. Close pack ice between the Kikkertaksoak Islands and the coast islands prevented the route being taken to Cape Harrigan, so passage was made through the inner sounds and Windy Tickle. Loose pans of ice, some of them very solid, were in all the bays and tickles, and the ship was continually under helm to avoid them. On her arrival at Cape Harrigan, loose pack ice was found from about a mile offshore to the Farmyard Islands but only drift ice inshore, so the captain hoped to round the cape; but fog came down, and the ship had to turn back to find anchorage behind Nunaksaluk Island.

It was not until 4 o'clock in the afternoon on the following day that the fog lifted sufficiently for the ship to get under way, but at once pack ice was seen almost in to the coast, with a lead of blue water about four miles out. There was a narrow lead between the pack and Cape Harrigan, through which the ship managed to pass, only about one hundred yards off the rocks but with a depth of sixteen fathoms of water. After she passed the cape, however, the course across to Wrecked Boat Island was entirely blocked, and the ship had to push her way through loose pack, butting the smaller pans end on, rising a little, and then breaking through with her weight. After she rounded the southwest end of Wrecked Boat Island, the pack was heavier, and the captain decided to stand inshore to pass inside Kutallik Island, the ship finally coming to anchor for the night in Davis Inlet.

On Wednesday morning the ship was under way by 4:30 and passed northeastward of Ukasiksalik Island, but the ice pack extended inside the Freestone Islands, and it was doubtful whether it was best to try and go through it or to take the inner channels. The inner route was selected, as Captain Clarke thought it likely that the ice would become worse towards Spracklins Island. The lead in the ice divided into two between Tunungayualuk and the mainland, and choosing one of these leads, the ship soon passed the wrong side of an island and had a rock awash ahead and had to bring up. The two motor sounding launches were lowered and with lead and line tried to find a passage, but with no result. So the ship turned back and tried the other lead, with the boats ahead of her running submarine sentries. A sentry is a simple device made of two boards about two feet long and six inches wide, secured at right angles, which, when towed from a boat, keeps a steady depth according to the length of the towing line. When the depth of water is less than that to which the sentry is set, it strikes the bottom and rises to the surface to give warning of shallow water. This device had been used by generations of surveyors for feeling their way in uncharted waters. The sentries soon tripped, and the boats then commenced a search for deeper water; it was not until 5 o'clock that they found a passage with six fathoms of water through it and the ship was able to proceed. The boat's sentries tripped again in the Narrows between the Tuktuinak Islands and Tunungayualuk. The ship anchored while they searched for a passage, which they eventually found, and then went on as far as Akpitok Island, where, at 10 o'clock at night, she brought up in eighteen fathoms between the island and the mainland.

By Thursday, rainy weather had set in. The ship weighed anchor at 4:30 in the morning and stood across to Dog Island through a lead in the ice. The existing chart was wrongly orientated here, but a channel was followed between Nochalik and Kikkertavak. With a boat running a sentry ahead, the ship passed through the Taktok Tickle and then northwestward for the channel between Palungitak and Paul Island, sometimes known as Pownal, where she came to anchor, and a boat was sent to sound out and buoy the channel through.

The area where the survey of the coast was to commence had now been reached, and all eyes on the bridge were alert to see any site ashore that would be suitable for the base line measurement. *Challenger* was about to be engaged in what might be termed a fundamental survey of an area where no geodetic positions or control had as yet been established. There are three basic essentials to such an original survey. First, a suitable stretch of land, at lease half a mile long, must be found and levelled. On this a base line is carefully measured with steel tapes to establish the scale of the survey. Next, the geographical position on the earth's surface of some point within the area of the survey must be fixed by taking star sights. The position of this point in relation to the base line can be found by measuring the angles of the triangle it forms with the base line. But the orientation of the survey still remains to be found. In other words, the direction on the earth's surface of the base line (or any other line on the survey) is still unknown, and this is established by measuring the angle between a line joining any two points and the sun (or a star), knowing the exact time at which this reading is taken. This is known as a true bearing between the points. Stations marked by flags on poles or by tripods boarded up or filled in with calico are then established, usually on the higher ground, throughout the area to be surveyed. By measuring the many horizontal angles contained by the numerous triangles thus established, the relative positions of all these points are found. The geographical position of each can be found by referring all stations to the position fixed by star sights and the position where the true bearing has been observed. This framework is known as the triangulation, and to it all the details of the survey, soundings, and topography are fixed as the work proceeds.

No really suitable place was found for a base on Paul Island, and so when the boats had sounded out the channel, the ship went through and came to anchor in Nain Bay at 9:30 at night, one week out from St John's. *Challenger's* arrival at Nain was popular with the few inhabitants of this little Hudson's Bay Company trading station, as she was the first steamer to get north through the ice that year, and the mail she brought was very welcome.[2] Nain lies well protected from seaward by numerous islands, the largest of which, South Aulatsivik, lies to the northeastward. A narrow channel, or "run," as it is called, divides this island from the mainland and leads to Port Manvers, about twenty miles northward. Local opinion seemed to agree that the ship could pass up inside this Port Manvers Run, although there were two "rattles" to be negotiated. (A rattle is a narrow channel through which the tidal stream flows so fast that an appreciable noise is made by the water.) Local opinion also agreed that a flat expanse of land lay on the south side of Port Manvers, which seemed suitable for the base measurement. There was no hope of reaching Port Manvers by the outside route for some weeks, pack ice being solid right into the coast of Aulatsivik. So on the morning of Saturday, 16 July, the ship steamed up the Port Manvers Run, experiencing no difficulty until the second rattle was reached about two miles from Port Manvers. Ice floes were drifting through the rattle on the tidal stream, so the ship was anchored and a boat sent to investigate. She located a navigable passage close to the eastern shore and later led the ship through, but as the point was rounded, it was seen that the whole western entrance to Port Manvers was jammed solid with ice, and there was nothing for it but to turn round and go back down the run to Satosoakuluk Island, about eight miles from Nain, which had been noted as a possible base measurement site.

The base site having been chosen, the survey went forward in earnest. On the evening before a day of surveying, the captain makes his plans, stating broadly in his surveying order book what work is to be done and which surveying officers are to do it. The first lieutenant then assesses how many men and what boats will be required for the various tasks, and then with the coxswain, he details the men for each party; these are not confined to any particular branch in the ship – seamen, stokers, the sick bay attendant, the officers' stewards, or even the ship's cooks may be sent on this work if they volunteer for it. To an outsider there appears to be a certain degree of chaos in the early stages of a survey; parties go off every morning after an early breakfast, taking with them their dinner, together with a quantity of gear, such as poles, flags, calico, rope, hammers, spades, mauls, bolt staves and axes, maps, charts, binoculars, and prismatic compasses.

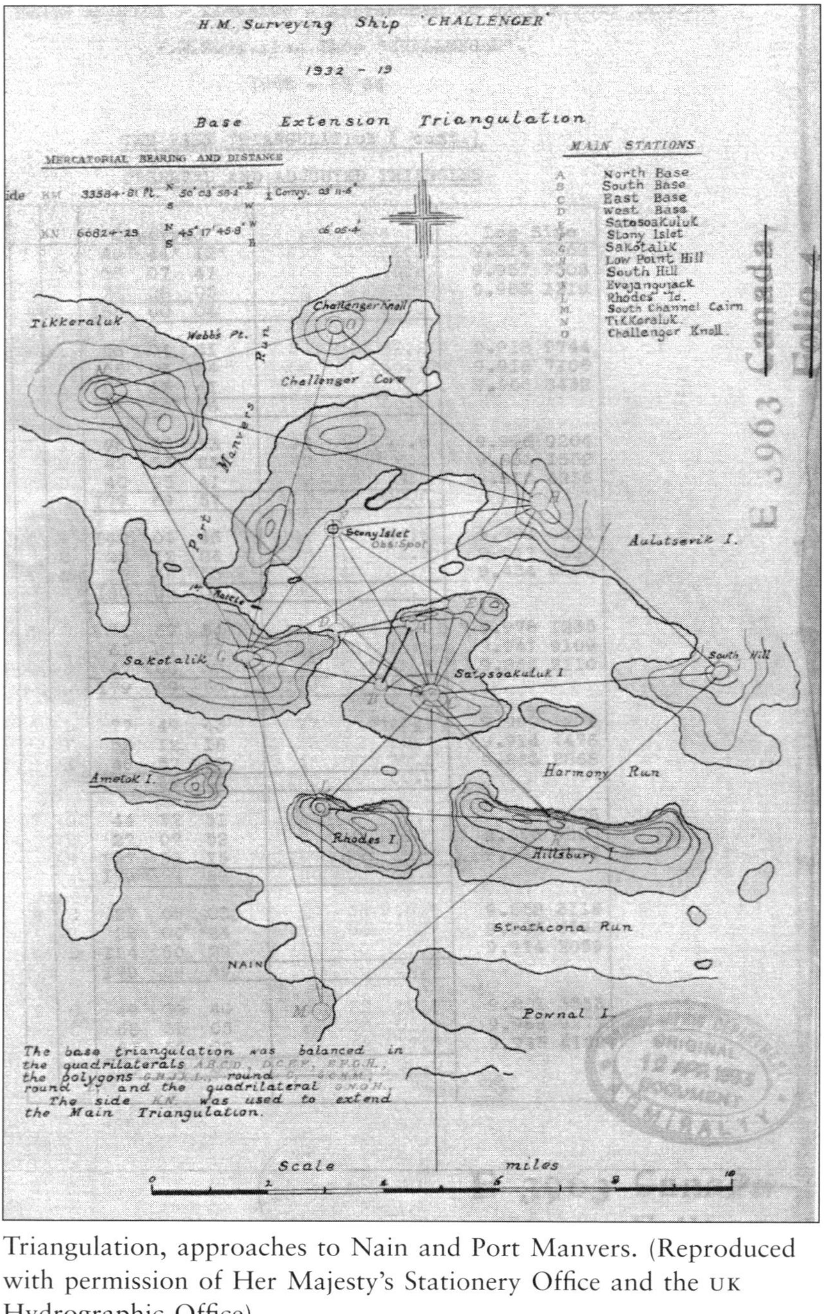


Triangulation, approaches to Nain and Port Manvers. (Reproduced with permission of Her Majesty's Stationery Office and the UK Hydrographic Office)

On this first day of the Labrador survey, a large party of surveying officers and men went ashore to prepare the base for measurement, while other parties were marking the proposed triangulation stations on the mountain tops, some of which are as much as three thousand feet high; such heights take a deal of climbing, starting from sea level laden with spars, ropes, and iron pegs for erecting the mark on the summit. The first few hundred feet of the mountains were clothed with spruce, and after that the climb became steep and rough, the mountain sides being almost entirely barren and even moss and lichens being scarce amongst the huge tumbled boulders. These same mountains would then have to be scaled again, with a theodolite on the back, to observe the horizontal angles between the other main stations and the marks that would by then have been set up along the coastline; and then perhaps again after that if the visibility closed down on the top, preventing observations.

The base measurement party had a formidable task ahead of it; part of the proposed base lay over stony, rough ground, while the remainder of the distance was across a swamp. It was the best that could be found in this rugged country. Three days' work was required before the hard ground had been levelled and an earth causeway built across the swamp to take the steel tapes. The measurement itself must be made with great care; even the temperature of the tapes and the tension upon them must be taken into account. It is a careful and unhurried process.

All surveying parties cursed the black flies and the mosquitoes that swarmed on the exposed parts of the face and body; head nets and gloves were essential, but if the head net was allowed to touch the face at any point, that part of the net would soon be black with flies, irritating damnably. Only when a strong wind was blowing was there any respite from these pests.

Rain fell heavily and steadily throughout these early days, and sometimes the whole base line area seemed to be under water. A portable galley had been rigged ashore to provide a good midday meal for the party, and the junior ship's cook had been established in a tent to do the cooking. He had never cooked under such conditions before, and soon he came out into the pouring rain carrying a little billycan: "Where do I go for water?" he asked a seaman who was drenched to the skin. "Water!" shouted the sailor. "Why, you just keeps on walking till your ruddy cap floats."

The prismatic astrolabe

Through August and September the triangulation work went ahead, the many horizontal angles being observed either with Watts theodolites or possibly with the new Tavistock theodolite, which Cooke, Troughton and Simms of York, England, had brought into production in 1930. Stony Islet was selected as the site for the observation spot, where a small camp party was established for making star observations with the 45° prism astrolabe, also made in small numbers by Cooke, Troughton and Simms in the 1920s. This instrument was set up and levelled on its small tripod, depending for its artificial horizon on a pool of mercury contained in a shallow, circular copper trough. A tubular compass attached to the instrument enabled the azimuth ring to be orientated to the nearest degree. Nightly star programs had been worked out using stars of 4.3 magnitude or brighter, which would appear between 25° and 65° from the meridian on reaching 45° altitude. Prisms within the astrolabe were so arranged that by operating a horizontal slide, the observer obtained six shots at each star. He recorded each shot by pressing his signal key, which punched the tape being moved forward by clockwork across the chronograph in the adjacent tent, on which time was also punched every second.

Through the long, dark nights the observer sat at the astrolabe. His assistant was in the tent operating the chronograph and alerting the observer as to when and in what bearing he should expect the next star. Sometime during the night, work passed while the assistant operated the radio to read a rhythmic time signal from Rugby in England. The data observed enabled the geographical coordinates of the observation spot to be found, and then, the position being known, a true bearing with reference to the sun was observed between two of the triangulation stations in the survey.

Two and sometimes three camp parties were established in the Port Manvers Run area and on the barren islands to seaward. The pack ice had cleared, and the ship was often under way among the islands, navigating day after day in uncharted waters. Apart from the day-to-day organization of the hill-marking and theodolite-observing teams, there was the welfare of those in camp to be thought of, replenishment of the camps with food and sometimes water on the more barren islands, and wood to be chopped on the mainland and carried to camps where no wood existed.

The Doctor Forbes Air Survey seaplane and a party in an attendant schooner arrived one day and provided welcome new faces; the seaplane also took a number of air photographs, which were of considerable assistance in the survey. Commander Wyatt was taken up for a reconnaissance flight over the area. The day of his flight was fine and clear, and he described the scene: the islands along the coast were almost literally innumerable, and the general appearance from aloft of the relation between land and water was that of pieces of a jigsaw puzzle scattered over a blue cloth. The country inland was exceedingly rough, well wooded in parts but broken up by many inlets of the sea, some running thirty miles or so inland, and with lakes and ponds scattered everywhere and at all levels. The immensity of the survey task was fully apparent from this height. Leaving a camp party at Nain, the ship hurried to Halifax at the end of August for fuel and provisions and then down north again, now using the easier outside route.

Back on the survey ground, the triangulation was carried seaward on the chain of islands, and with all the officers except the boatswain either away in camp or with daily survey parties, the captain was alone all day on the bridge as the ship moved from one island to another, picking up parties from one point so that they could be landed elsewhere to observe there. The ship moved amongst a complexity of uncharted islands, rocks and shoals, feeling her way, constantly under helm, and sometimes, with breakers ahead, having to come astern or turn round and try another passage.

A secure anchorage was found at Ford Harbour, at the east end of Paul Island, and thither the ship often made her way back as dusk was falling, only to be under way again at the first light of dawn. On Friday, 23 September, she weighed and left Ford Harbour at daylight and steamed northwards at nine knots to land a theodolite-observing party on an island about eight miles distant. She was taking a route that she had followed in safety twice previously and where no dangers had been located by echo sounding or by the man looking out from the crow's nest (such a lookout was always posted in these waters). The captain was alone on the bridge except for the rating at the echo-sounding machine, who called out the soundings every twenty seconds. The soundings were between fifty and thirty fathoms, but suddenly a sounding of twenty-one fathoms was called, and the water then shoaled so rapidly that the echo was momentarily lost, and before another sounding could be called, the ship shuddered as she struck a rocky shoal. The engines were stopped and put to full astern with no effect. Abreast the bridge the ship was hard and fast, while at her bows and stern soundings showed deep water; truly, this was a pinnacle rock. It was indeed a serious predicament, the ship firmly aground in a remote part of the Labrador coast with the nearest help over a thousand miles away and the amount of damage done to the ship's hull difficult to assess.

A boat having sounded round the ship, it was decided to attempt to get the ship off stern first at high water, which was due in about four and a half hours' time and might give about two feet of water more than when the ship grounded. Time was short and work commenced at once. First, the damage was inspected as far as this was possible inside the ship, and then wooden shores were placed against all the adjacent bulkheads, the damaged compartments themselves being sealed off by their own watertight doors. It appeared that all the forward fuel-oil and fresh-water tanks were leaking and that the large provision room and canteen store were flooded with fuel oil and water, and further, that there was a slight leak in the forward end of the boiler room, which was situated abaft the provision room. This damage must be localized as far as possible so that when the ship floated off the reef, the minimum risk of further damage might be incurred.

Meanwhile, the two ship's anchors were lowered below the waterline and slung beneath boats; this was no mean task, as each anchor weighed twenty-eight hundredweight. These anchors were disconnected from their cables, and when wire hawsers had been secured to them, they were laid out astern of the ship and the hawsers brought to the steam trawl winch on the quarterdeck. An amusing incident happened at this point, for when the huge anchors were slipped from beneath the boats, the latter regained their full buoyancy with such force that a member of one of the crews was catapulted into the icy water. A smaller kedge anchor was similarly laid out ahead of the ship to steady her.

Just before high water, the anchor cables, now without their anchors, were lowered onto the seabed, fourteen tons of fresh water and fifteen tons of fuel were pumped overboard from tanks in the forward part of the ship, and all the crew mustered aft, so that the ship was considerably lightened forward. At about high water the stern hawsers leading to the two bower anchors, which had been laid out astern, were hove upon and the engines put to full speed astern. It was a tense moment as ever so slowly the ship began to move and very gently came off the rock. The first part of getting the ship back to port was achieved, but there were many more difficulties to overcome yet. A hum of eager conversation on the quarterdeck was soon quelled by the coxswain detailing the men for the task of recovering the anchors. The hawsers running to the bower anchors had now to be led to the forward winch and these anchors hauled up into the hawse pipes once again and connected to the cables. The kedge anchor had also to be recovered before the ship returned to Ford Harbour, so that divers could determine the full extent of the damage; meanwhile, the pumps were just holding their own with the water in the damaged compartments.

On the next day, Saturday, a strong easterly wind was blowing into the anchorage, and it was late in the afternoon before the two ship's divers could be sent down. The whole of Sunday was also spent in diving, while the ship's boats made long journeys to recover the three camp parties, who were soon alive with bustle as they took down the tents and packed the bedding, provisions, surveying equipment, and a hundred other items that make up the detached survey camp. The divers had an extremely cold job, and they could work only in short spells before warming up onboard and preparing to go down again; but encouraged by the boatswain and the small party of men who were working the air pump and attending on them, they gradually built up a picture of the damage to the ship's hull, and in addition, they were able to plug six holes from which rivets were now missing. This repair work allowed the pumps to reduce considerably the amount of water in the damaged compartments.

At daylight on Monday, 26 September, the ship was as ready as she could be to sail for the south, and it was now necessary to take the inshore route again, at least until it was certain that the plugged rivets were holding satisfactorily. To encounter heavy weather such as could be expected in the open sea might cause further damage, particularly to the forward bulkhead of the boiler room, which might be flooded with the most serious results. Both *Challenger's* boilers were in the same boiler room, fitted side by side.

So once more the boats swept ahead of the ship with the submarine sentries for two long days as the ship sailed southwards through the islands to Cape Harrigan. Halifax seemed a very long way away to those who were anxiously watching the damage and nursing the vessel every mile of the way. The British naval commander-in-chief of the West Indies Station had been kept informed of the ship's plight, and on Wednesday she met with the naval sloop *Heliotrope*, which had come north to stand by.

The glass began to fall, and the wind was freshening from the south; such weather might soon increase the damage, so the two ships anchored in Domino Run for shelter. There they remained weather-bound until Friday, when a lull permitted them to go on to York Harbour in the Bay of Islands, on the west coast of Newfoundland, where again shelter had to be taken. Never more anxiously had those on board watched the barometer and the freshening wind. After a false start on Sunday, when the swell in the open sea forced them to return, they eventually sailed on Monday and reached Halifax on Wednesday without further mishap. Here they were received with generous assistance afforded by the Royal Canadian Navy, and the ship was docked a few days later for extensive repairs to be carried out by Halifax Shipyards Ltd. It was not until 18 November that she was ready to sail home to Portsmouth.

The Lords of the Admiralty considered this case of grounding and decided that no blame was attributable to anyone for striking an uncharted pinnacle rock in such difficult waters; furthermore, they considered that the commanding officer had acted in a seamanlike manner in getting his ship afloat and carrying her to Halifax, and they commended Mr H. Good, the engineer officer, and the divers for their fine work. The Hydrographic Office placed "Challenger Rock" on the charts of the Labrador Coast.[3]

Discussions at the Hydrographic Office during *Challenger's* winter lie-up led to a decision that, because of the brevity of the summer surveying season in Labrador imposed by the ice conditions, a winter survey party should be established at Nain in November. It seemed that such a party might expect to make considerable progress in both triangulation and coastlining work; with the sea frozen over, the theodolite teams would be able to move across from one island to another by sled, while coastlining could be progressed by sledding along the edge of the frozen water.

Although the ship reached the coast of Labrador again on 21 July 1933, only about a week later than the previous year, the ice situation was much clearer this year, and the passage northwards simpler. The ship anchored in Davis Inlet so that the captain might visit the senior Hudson's Bay Company trader there to make early arrangements for winter clothing and other supplies needed for the proposed winter party and required to be sent north to Nain. The next day the ship again reached Ford Harbour, where nine months earlier she had been licking her wounds after her encounter with Challenger Rock. The poles marking the triangulation stations, which had been left behind on the ship's hurried departure the previous year, were still standing and could be seen on the hilltops. Next morning parties were away putting new flags on these poles, from which the winds of winter had removed every trace of the old.

A conference was held on board at Nain to which Mr Grubb, the missionary, and Mr Clarke, of the HBC store, were invited so that arrangements could be made in good time for the winter party, who were to have their headquarters in Nain. It was decided that the party should live in the small, unused wooden hospital, which appeared sufficiently large to accommodate them and to leave room for a chartroom for plotting the survey work. Arrangements were made to start collecting the dogs that would be required for hauling the sleds, or komatiks, which were of the simplest construction, being formed of two long fore-and-aft wooden runners shod with iron, with cross boards secured between them on the top along their entire length to form a platform.

The marks having been, for the most part, fixed the previous season, the ship and boats were able to press on with the sounding work, while parties landed daily to delineate the coastline. All ship and boat sounding work was fixed by the long-established sextant and station pointer method. At the end of September the ship, leaving a large camp party with both sounding boats in a landlocked bay south of Nain known as Kauk Harbour, sailed south to Halifax for fuel and the large stocks of provisions needed for the winter party. By 12 October, in the first flurry of winter snow, these stores were being disembarked at Nain.

From now onwards, the survey work was much hampered by snow and gales; on 2 November the temperatures was down to −11°C, and spray was freezing as it fell on the canvas canopies of the boats, which were away placing secondary marks along the coastline for the use of the winter party. While one of the boats was being hoisted on this day, a brand new fall parted in one of the blocks; as the whole weight of the boats is taken by these ropes, it was fortunate that no one was hurt and that no damage was done. When the fall was examined, it was found

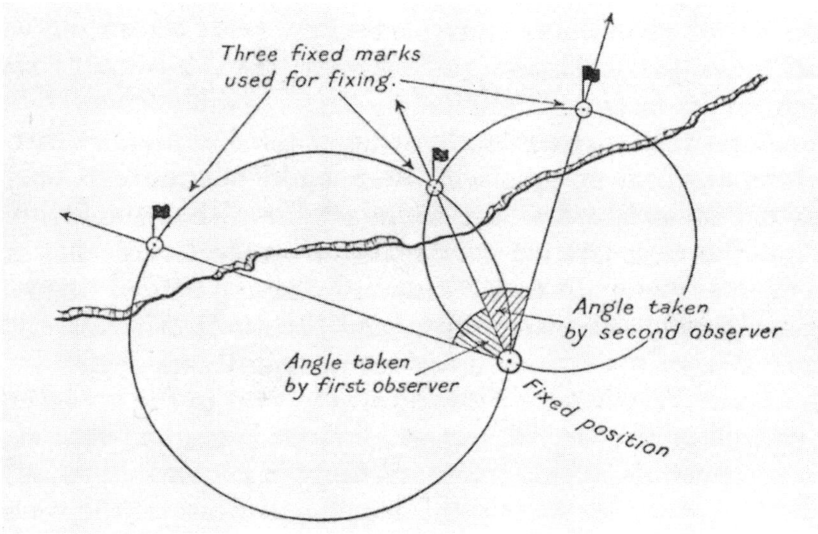


Station pointer fixing

that, because of the low temperature, the tarred fibres had become frozen and brittle, so that, when bent, the rope snapped like a stick. So from now on, falls had to be unrove after a boat was hoisted or lowered and stowed below until required again, a most unpleasant nuisance in such conditions.

On 9 November strong westerly winds were blowing the snow off the hills like smoke, and next day the ship proceeded to Nain to establish the winter party. The temperature was down to −10°C, and ice was forming around the ship's waterline. All the stores were ashore and stowed in the old hospital at Nain by the evening of 14 November, and the party said farewell and went onshore. The next day the glass fell to 956 millibars, but there was no wind until the afternoon, when it began to blow from the southwest and gradually increased to a gale at night. The temperature fell to −20°C on board, and when an attempt was made to veer more cable, it was found that the windlass was frozen solid. On the morning of 16 November the sea was "smoking" prior to freezing over; it seemed high time to be away. A fire had to be lit under the windlass before the cable could be worked, but once that was freed, the ship was soon heading for the open sea in a fresh gale and a snowstorm.

As the ship steamed slowly out of the bay and sailed for home, members of the winter party turned to their immediate tasks. The hospital was scrubbed out, and shelves and hooks were erected to take the stores and the winter clothing; the chartroom had to be rigged and provided with a table, chairs, and racks to take the instruments; and the most important work of hauling in the firewood and stacking it in readiness for use had to be started at once. Meanwhile, the ice began to form in the bay, at first very thin and easily broken up by the slightest wind, but on 22 November news came in that the Port Manvers Run was completely frozen over except for the rattles and that people were able to walk on the ice there. Sledding would soon begin.

The population of Nain consisted of a number of settlers of European extraction and a few Inuit families, some living in the settlement and others in huts, built as fishing and hunting camps and tucked away in odd corners, at the head of an arm of the sea or close under a sheltering spruce coppice. The winter party consisted of Lieutenant-Commander "Buck" Baker in charge, with Lieutenant Dennis Deane as his assistant surveyor, a cheerful and amusing person; then there was "Doc" Bingham, who was a surgeon lieutenant-commander. The ratings consisted of Petty Officer Stevenson, Leading Seaman Hampson, Able Seamen Marshall and Marlowe, and lastly, Officer's Steward Holgate. They had all volunteered for this winter in the north, but only Bingham had experience of dog-team driving, of the use of snowshoes, and of living under Arctic conditions. There was much to be learned, and the party were soon making their first floundering steps on snowshoes, while they tried out the winter clothing, the windproofs, the sealskin boots, and the sleeping bags. Henry Voisey, one of the local settlers, was taken on as camp servant. He was a good man with dogs and had his own team, and he knew the country well.

Baker at once decided that he should be ready at any time to send out a party capable of living unsupported for at least seven days, so a number of "camp boxes" were made up and kept in readiness to be loaded onto a komatik. These boxes contained saucepans, a frying pan, a can opener, a primus stove, knives and forks, and so on, sufficient for two persons, while a "ration box," which would also have to be taken, contained concentrated rations sufficient for two persons for seven days and included such items as six pounds margarine, six pounds pemmican, porridge oats, plasmon biscuits, pea flour, and cocoa.

Baker decided to go with Bingham and a dog team to establish a camp for a week, so that they might judge what the conditions would be in a survey camp and what snags they might face. So on 29 November, having loaded the komatik with a camp box, a ration box, a tent, and sleeping bags, they set out on their first sledding journey, with Baker's and Bingham's teams harnessed together, to make camp, after four hours' travelling, on the shore of a saltwater lake. And here for a week, shooting ptarmigan for winter use and fishing through a hole in the ice for cod to feed the dogs, Baker and Bingham acquainted themselves with the difficulties and the conditions that would be experienced by those occupying the survey camps during the long winter months ahead.

The tents were of a new Arctic pattern, having a double skin and a ventilator in the outer canvas that could be opened by a cord from inside; but when the Primus stove was out, it was found that hoar frost formed both on the inside and between the two skins, making the tent very damp when it thawed on the Primus being relit. So tubes were sewn from the inside to the exterior ventilator, thus reducing considerably the amount of hoar frost that formed whenever the temperature in the tent was allowed to fall. By spreading deerskin rugs between the sleeping bags and the ground sheet, the bags were kept moderately dry, and once the sleeper had overcome the sense of suffocation experienced when using the hood, he was able to sleep warmly and in comfort, even when the temperature in the tent was down to −18°C, as it often was.

Driving the dog teams had also to be practised. A komatik is normally pulled by seven or so dogs, which are harnessed by a single bridle that, at some distance ahead of the komatik, has short extensions or traces leading to the dogs on either side of it. From experience it was found that traces and bridles made of sisal rope of three-quarter-inch circumference brought from the ship were far easier to manipulate and were also stronger than the sealskin line normally used for this purpose. Little difficulty in the actual driving of the dogs was experienced in the early days, although much practice was required to get used to handling the forty-five-foot thonged sealskin whips, which, when well manipulated, can be used to flick a delinquent dog in exactly the right place from a position on or behind the komatik; the inexperienced driver often ends up by giving his own cold face a savage cut with the end of the thong.

On 12 December the advance gear for a survey camp – tents, rations, hop poles, and flags for survey marks – was hauled by komatiks to the first site selected, at the mouth of a small bay on the eastern shore of Satosoakkuluk, about eight miles from Nain. This gear was cached among the trees at the site, and the teams returned to base. It was intended that the survey party would come out on the following day, but as was to happen so often throughout the coming winter, the barometer fell during the night, and it came on to blow hard from the westward. The visibility on such occasions is cut down to a few hundred yards, which, combined with the bitterness of the wind, makes travel impossible. So the dogs huddled on the lee side of the old hospital, and the men lay up all that day. But they hauled out to the camp site next day and having set up the camp, carried out a considerable amount of triangulation in the area. A week later camp was struck and the teams had a hard trip back to Nain, as the komatik was heavily laden and one dog was sick, leaving a team of six dogs to haul across the bay against a strong headwind blowing from the west. As soon as the base was reached, the party were helped by a number of Inuit to unharness the dogs and to unload the komatik. This was a local custom: any komatik coming into Nain was always assisted on the last part of the journey by anyone who happened to be around, and when all the gear had been carried into the house, the helpers melted away, expecting no thanks for this courtesy, which always made the return to Nain seem like a homecoming and was very welcome after hours of battling with the elements.

Preparations for the second survey camp were made on 3 January, and the next morning a small party led by Baker was away to establish and occupy the camp. His spirited private diary describes the rigours of the surveying work and the day-to-day difficulties of sledging and camping.

4th January. It looked as though it was going to be a really good day as the wind dropped away altogether by 0830, but came in puffs from all points of the compass. It finally steadied in the S.W. and by the time all the gear was lashed to the komatiks by 1000, it was getting up. We started off and the wind increased to about force 6 before we had got very far, and fairly blew us along. We got to the camp site by noon and pitched camp in the same place as last time. During our absence, however, a spring must have burst its way through the snow and ice as Marshall fell through into about a foot of fresh water about 15 yards from our snow shelter. Henry started building a snow house for us to use as a survey office but made it rather big and with not enough tumble home so that it was impossible to put the roofing on "à la Eskimo." I made him pack up about 1530 and sent him back to Nain as I wanted him to get back before dark. We settled down comfortably and had all the fresh water we wanted for cooking and drinking without the bother of having to melt snow.

5th January. A cold day with wind from the N.W. but not a strong one. We all went away coastlining in different directions. Personally I don't think that I have ever had such a foul day, as the wind, although not strong, was darned cold and going against it simply froze one's head and face. I went up wind and coastlined down wind but it was pretty grim as every time one stopped to fix and "shoot up," one got beastly cold, the sextant telescope clouded over and this immediately froze and had to be cleared off with the point of a pencil. Taking angles up wind was very trying as it made my eyes water and this, of course, froze too. The temperature was −17 degrees all day. I got back to camp just about ready to commit murder at the slightest provocation. As the snow house was not completed, Dennis and I had to plot and "ink-in" in the tent and this was about the most trying thing I have done for a long time. Only one could work at a time and he had to sit on the ration box with the plane table on his knees and tilt it downwards away from him to get enough light to see by. Our eyes ached like hell by the time we finished and we both vowed we would not try it again. Altogether somewhat of a grim day.

6th January. It was snowing to start with so we finished building the snow house, and set off on our coastlining as soon as it cleared up. We used the snow house in the evening for plotting and inking-in and it was simply marvellous compared to a tent. The plotting table consisted of a large snow block laid on top of a beaten down pile of snow in the middle and the light was supplied by two candles lashed to broom handles and stuck in the snow either side of the table. The light given was truly amazing and made work almost a pleasure! The primus stove kept it fairly warm and we slung the bottle of ink on a lanyard to thaw it and keep it thawed. Even so we had to hold the ink over the primus flame every minute or so as the ink froze on the pen while inking-in. We finished the great work by about 1830, and on going out noticed that the spring was boring up almost into the igloo although this was built on the side of a small hill. After supper I went out to do the usual chores of filling up pots and pans and so on and went up and had a look at the igloo and found the water had risen considerably and also that it had broken out at the back and was threatening the tents. I tumbled the crowd out and we cut a channel to drain the igloo and cut through into a small river which welled up very strongly. I decided to shift camp and we made a quick job of it, shifting it about 50 yards away in a lee of a small hill. The whole camp was shifted and settled down again with all primus stoves roaring in three-quarters of an hour. The lads were very cheerful about it and picked the tents up bodily and carried them to the new site, singing some Salvation Army song that is usually sung when the latter carry large banners about. I did not feel too happy about the snow house but, as we had tapped the stream below it, it did not look as though anything would occur.

7th January. It was blowing hard and continued to do so until about 1130 when the wind appeared to ease somewhat. We all had a quick "mug up" and got away by 1230. It was soon apparent, however, that it was only a temporary ease up, as by the time I had cleared the land it was blowing hard again. Standing up in the komatik with the wind behind me it was strong enough to blow the komatik along and keep the bridle and traces slack over hard windblown patches of ice. By the time we had got to the east end of Kruger Island, I could see that we were in for a buster as the snow was being blown off the tops of the hills in Nain Bay. I rounded the point and started to come back the other side of the island, as I thought we might have a lee of sorts. I was mistaken, however, for very soon it was blowing so hard and with so much "drift" that I could not see my dogs. It was evident that we were not going to get back stemming that wind and drift so I got ahead of the team and led them up over the top of the hill 180 feet high. They followed like lambs and pulled the komatik up like little Trojans. We careered down the other side and found it was much better as far as drifting snow was concerned, but the wind was so strong that we had to lean against it. We got back to camp by 1530, and on getting up to the camp saw two of Dennis' dogs there and so thought he must have got back before me. I could not see his komatik anywhere, though, and realised that one of his dogs had slipped its trace and run back to camp and that he was out in the gale with only four dogs. I went over the crest of the hill to look for him and told Marlowe not to unharness my team as, if he was a long way off, I would go out with my team and bring him back. I met him, however, just coming up the hill and gave him a hand home. He was breathing fire and murder about his leader, Lively, as he thought he had gone back to Nain again, as he had done on Friday. He was very surprised to find him in camp and forgave him. The other dog, Frank, I knew he had left in camp as it was sore under the forelegs. We were all glad to get back and get into our tents out of the wind.

8th January. It was still blowing when we turned out so I decided to wait and see what the weather was going to do, as the barometer was going down and it looked as though anything might happen. On going over to the snow house we found to our horror that the water had bored up inside and my guns, harnesses, plane tables, cartridges and other odds and ends were standing in about six inches of half-frozen water. Dennis crawled in and chipped all the gear out and passed it out. We shifted all the surveying gear to a little dip on the opposite bank of the pond that we were now camped on. It looked as though the snow was good enough to build another snow house so we all set to work and had one up by lunch time. It was made more or less in the correct way and was roofed "à la Eskimo" and was a great success, thanks chiefly to Dennis who did most of the building. The wind was dying away so we had a hurried lunch and got away by about 1300. We had a good afternoon surveying as the wind died away completely. I did not get back till dark, and we plotted and inked-in in the igloo again and by the time I came out it was snowing. The barometer was falling fast and looked as though we should have bad weather in the near future so we decided that, if it was not surveyable weather on the morrow, we would pack up and go back to the base.

So work continued throughout the winter. Easter not only provided a change in the daily routine, but it also marked a change in the seasonal calendar. Good Friday fell on the 30 March, and while members of the party were attending the church service, the weather turned suddenly mild with heavy rain, and soon there was much slush underfoot. By Sunday, when the band climbed to the church roof to give a concert, all the snow had gone from it. Spring was coming to Labrador. The snowstorms still came in the following weeks, but the snow was soft and wet, making poor going for sledding; the komatiks proceeded slowly, and the men waded knee-deep beside them.

During the long winter months, those who did little travelling spoke of the spring with delight as one would speak of the coming of spring during the dark months of an English winter, but to men such as Willie and Joe Ford, who spent their lives hunting and trapping, the spring held no illusions, and they said so. Willie Ford said he loathed the spring; the going was nearly always bad, and one suffered from a blistered face and cracked lips because the heat from the sun was reflected by the snow. Even when snow was absent, there was water about a foot deep covering the ice, through which the dogs had to wade. Baker describes in his diary travelling in the spring snow: "The speed of travel was the speed of the dogs … as the poor beasties were belly deep the whole time and they had to lift their feet out of each step as the snow was too cloggy for them to push through. Walking alongside was no joy, plod, plod, the whole time and nearly always knee-deep … I found it was 5 p.m. and we had taken 7 hours to do 10 miles."

Springtime travel

In early May it became necessary to lash a boat on top of each komatik, and in this all the camp and surveying gear was stowed, so that in the event of breaking through or having to cross deep water lying on the sea ice, the komatik and the "flat," as the boat was called, would keep afloat. It was a peculiar experience for the travellers to sit in the bows of a boat, the surrounding water being above the top of the komatik, while they looked ahead at the wading dogs towing this amphibian. So with their faces covered in Vaseline, their clothes sodden, and their tempers frayed, the party worked on through the spring, more coastline being plotted and more topography mapped, as the outline picture of the coast grew and grew in readiness for the ship's return, when she would complete the work by sounding out the runs, the bays, and the deepwater channels leading to the open sea.

On 22 May the last survey camp of the winter was established, this time in a trapper's hut on Bridges Run. The brooks and the runs were opening up, and every day detours to avoid open water became longer and longer. On Monday, 4 June, it was decided that komatik travel was no longer economical, and the final return to Base Camp was ordered. Small paying-off pennants, such as are hoisted by a ship at the end of a commission,

"Buck" Baker and Dennis Deane going out to meet *Challenger*, 23 July 1934

were hoisted in the bows of the two flats, and the drivers beat frying pans to encourage the dogs. They left Bridges Run shack at 10 o'clock, and the going was surprisingly good after an overnight frost, although the komatiks were afloat from time to time. Racing down the bay to Nain with the pennants flying and the pans beating, the two komatiks made a brave sight to the watchers in the settlement. The long sledding season was over, and the next few days were spent in disposing of the dogs to their original owners or those who wished to buy them.

In July the same year the Grenadines survey, on which *Challenger* had been engaged in the meantime, was broken off, and the ship headed once again for Labrador, calling at Halifax for fuel and reaching Nain on 23 July 1934. There was less ice on the coast than in former years, and the passage north was comparatively simple. Baker and Deane were at their supper on this evening when someone rushed in and shouted something unintelligible, which, on investigation, was found to be news that *Challenger* had been sighted. She came to anchor at 7:15 p.m. off Nain. Baker and Deane were out in their kayak waiting for her before she anchored, and alongside and on board the moment she let go. There was a great reunion party, but many faces were new to the two, who had left *Challenger* nearly nine months before. It was 1 o'clock in the morning when the kayak left for the shore.

*Challenger* completed the survey by sounding out a good approach channel to Nain from seaward from the vicinity of the Hen and Chicken Islands and by sounding the coastal passages southward from Nain. She carried out a running survey through the Port Manvers Run and sounded out a narrow route southwards from Port Manvers through the labyrinth of islands that had been mapped by the winter party outside Aulatsivik.

The latter part of the season was spent doing a survey of Cartwright Harbour and approaches, both boats being left in camp there to progress the work while the ship visited Quebec City for fuel and stores, this being a pleasant change from Halifax, which the ship's company now knew so well. On 15 November, with the survey of Cartwright complete, *Challenger* sailed away from Labrador for the last time and reached Portsmouth on 24 November 1934.

7

# *Hydrographic Studies of Russia's Northern Oceans, 1900–1940*

VLADIMIR SERGEEVICH SOBOLEV
TRANSLATED BY GEORGE BOLOTENKO

Hydrographic studies of the northern oceans had great state significance for the Russian government. The northern littoral from Murmansk to Dezhneva Cape is washed by the White, Barents, Kara, Eastern Siberian, and Chukchi Seas; it constitutes a coastline that, including the shores of islands located in this region, approximates forty thousand kilometres. The harsh nature, severe climate, and heavy ice conditions of these reaches create unique difficulties for shipping and for the conduct of hydrographic work. At the same time, the northern littoral is for Russia its open egress into the world ocean, and its northern seas constitute the sole water route connecting the country with the measureless expanses of its eastern territories. This chapter looks first at the hydrographic administrative structure and methodology and then reviews the hydrographic achievements in the various geographic areas of the northern seas.

The beginning of the twentieth century brought many new demands for Russia with respect to the direction, scope, and quality of hydrographic work in the north. Some of the major reasons for this changed situation were the appearance of naval and merchant ships with deep draughts and greater speed; the noticeable and steady increase in cargo deliveries and the expanding activity of the hunting (and fishing) fleet; the need to ensure the safety of submarines and the ever-increasing complexity

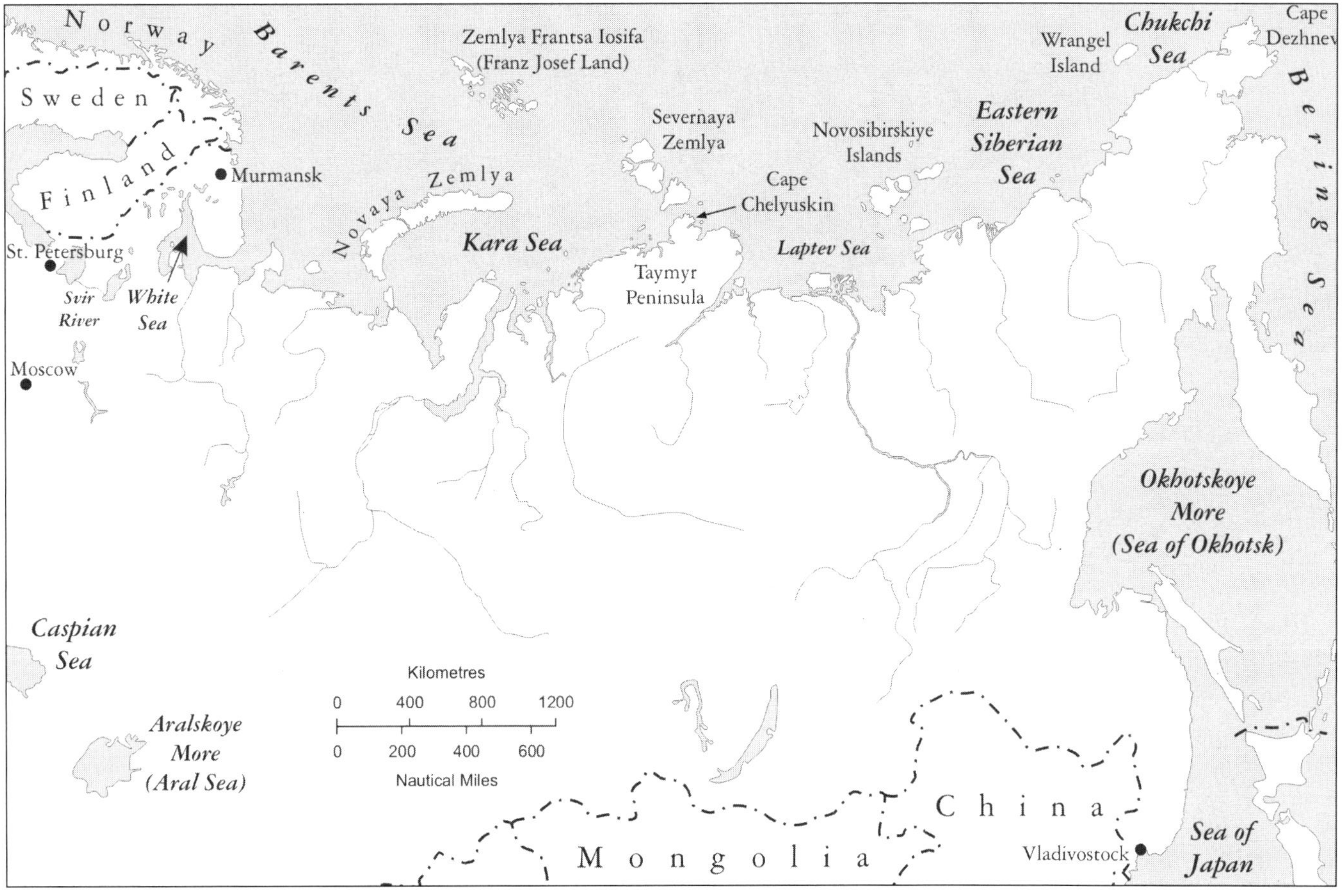


Russia's northern coastline (Map: Chris Johnson)

of employing mines and torpedoes in this area; and looming difficulties in the further exploitation of the natural resources of the north and of Siberia. One should stress here that all these developments required a response that was rendered much more complicated by the massive scales of the work to be done. Not surprisingly, the organization of the hydrographic service changed over the period under review here.

From 1900 to 1917 the hydrographic service was under the control of the Russian naval ministry, constituting one of the responsibilities of an independent unit of that ministry. From 1885 onwards, this unit was called the "Chief Hydrographic Administration." In 1891 a Cartographic Section was established within this unit, and in 1897 it began to publish sea charts printed from aluminum plates. Further, in 1902, a Photographic Unit was established within the Chief Hydrographic Administration, which successfully undertook the work of reprinting foreign sea charts.[1]

In 1907–13 General A.I. Vil'kitskii, a talented individual, well educated and an able organizer, ran the Hydrographic Administration. He managed to double his budget allocation for hydrographic work, as well as substantially improving the material resources at hand, which expanded the prospects for the development of hydrography. Subsequently, in 1910 he worked out a thirty-year plan for hydrographic work, as well as a ten-year plan for the erection of beacons.

To ensure a supply of qualified personnel, a Corps of Hydrographers and a Navigational Officers Class, both under the Hydrographic Administration, were established in 1912, but further development of the hydrographic service was interrupted by the First World War. It is worth noting that, in the prerevolutionary period, actual hydrographic activities at sea were managed by various units scattered all over geographically, not unified into a single structure – units such as the administration of navigational lights and pilotage, observatories and hydrometeorological stations, surveys and discrete expeditions, instrument workshops, and other units.[2] Naturally, the lack of coordination did not facilitate normal development of the service.

The post-revolutionary period saw reforms to both the naval and hydrographic services. In June 1918 a *prikaz* (edict) to the fleet instituted a new organizational service structure. Each sea would have its own "Administration for Ensuring Safe Navigation" (UBEKO). A significant aspect of this reorganization was that each UBEKO began to consolidate all the formerly disparate hydrographic units for each sea (lighthouses, weather stations, expeditions, workshops, and so on). In 1918 a new unit, Cartography, was established in the Admiralty at St Petersburg, which unified under its control all the former cartographic offices and began to manage all aspects of work in the publication of sea charts. At the end of the Civil War, it became possible to renew the training of hydrographic specialists. Four were trained in "Separate Hydrographic Classes" from 1921, and the following year the Naval Hydrographic College was established. From 1926 onward, hydrographers were trained in the M.V. Frunze Naval College.

The second half of the 1930s saw the UBEKOs reformed into the Hydrographic Sections of the Fleet, one each in the Baltic, Black, Northern, and Caspian Seas, as well as in the Pacific Ocean. Additionally, after the creation of the Chief Administration of the Northern Sea Route in 1933, that unit received its own hydrographic service. It is worth noting that right into the mid-1930s the hydrographic service, in the performance of its work, utilized, in the main, old vessels from the czarist period. Only in 1936 did the first specialized, newly constructed vessels appear.

At the beginning of the twentieth century, the methods employed in hydrographic work were marine surveys and the compilation of brief descriptions of the area surveyed. The geodesic bases of such descriptions were primarily local geodesic systems, distinct astronomical points, and at times, disparate and disconnected geodesic networks.[3] It should be noted that at this period, topographical surveys of coastal areas were far from complete; hence sea charts were full of inaccuracies and at variance from the areas represented.

During the early years of the Soviet period, the hydrographic service received extremely limited budgetary allocations; thus it had a weak material-technological base, and the capabilities of its expeditionary units were exceedingly limited. Because of these deficiencies, systematic investigation did not go beyond the preparation of a sea survey.[4] For example, in 1924 the Chief Hydrographic Administration received something in the area of only 300,000 rubles, which was intended to cover all its expenses.

According to this approved budgetary allocation, the hydrographic service received financial support for the following activities in the amounts indicated: 98,000 rubles for the maintenance of lighthouses and radio stations; 50,000 rubles for the maintenance of buoys, beacons, and shore markers; 70,000 rubles for economic and operational expenses; 60,000 rubles for surveys and soundings; and 27,000 rubles for the preparation of charts and instruments.[5] To give some sense of these amounts in today's terms, according to the official conversion rate noted by Gosplan in 1924, the Soviet *chervonets* (i.e., 10-ruble note) bought US$5.00.[6] Thus, to cover all the expenditures of the hydrographic service of the USSR, the government allocated annually a sum equal to roughly US$150,000.

At the beginning of the 1930s, Soviet hydrographers gradually began to develop systematic surveys based on the following principles:

- the starting point for all work was triangulation, where a base line of the triangle was set along the shore (the permissible error in the triangles was not to exceed six inches);
- these lines were verified by the measurement of the bases and Laplace points not more than three hundred kilometres apart;
- the usual scale of measurement was 1:50,000;
- the coastal and sea soundings allowed for error in depth sounding of not more than 2 per cent; and
- sounding work was done from ships' boats using the lead, while ships built for hydrographic work had echo sounders aboard.[7]

During the 1930s geodesic work was done largely with Wild and Zeiss instruments, imported from Germany. Towards the end of the 1930s, other contemporary methods of work were introduced, such as the use of photogrammetric methods to survey the shoreline from ships, new receivers for equalizers and other equalization techniques, and the interpretation of hydrographic objects from air photos, and other methods.[8]

Notably, from the beginning of the twentieth century, there was an annual increase, however gradual at times, in the extent of hydrographic work done in Russia, and by 1913 the area surveyed totalled 9,700 square kilometres (including the surface

Table 7.1

| *Year* | *Area surveyed (in square kilometres)* |
|---|---|
| 1925 | 1,789 |
| 1928 | 2,951 |
| 1931 | 13,495 |
| 1934 | 61,687 |
| 1937 | 123,200 |

SOURCE: *Zapiski po gidrografii*, 1938, 1:21.

area of surveys completed, soundings, and trawling). During the first years of Soviet power, little hydrographic work was accomplished, for obvious reasons. But the period from 1925 onwards was characterized by a marked, ever-increasing expansion, as illustrated in table 7.1.

Attention should turn now to the hydrographic studies of Russia's northern seas over the period 1900–40. The region is normally divided into four geographic areas: the White Sea, the Barents Sea, the Novaya Zemlya Islands, and the Northern Sea Route. At the beginning of the twentieth century, current knowledge about the White Sea failed to satisfy the growing needs of navigation in this area. The end of the nineteenth century had seen the creation of the Separate Survey of the White Sea to carry out the necessary work. However, the survey received very negligible technical, financial, and human resources. It comprised only seven to ten people actually employed in the work of surveying; up to 1896 the survey did not have its own ship, and for sounding purposes, it had to borrow a small vessel from the Administration of Lights and Pilotage of the White Sea.[9]

In the main, the work of the survey centred on Oneshskaya Bay, but even here soundings were done only for narrow strips along the main channels, and plane table surveying was performed only for narrow strips along the shores. The surveying of discrete sections was based on local triangulations, together with the fixing of astronomical points. Amongst the leaders of these projects, often completed under difficult climatic conditions, were Lieutenant-Colonel K.A. Miakishev, E.V. Maidel, M.E. Zhdanko, and N.M. Deploranskii.

The survey's report of its activity during the summer months of 1901 notes that the following was accomplished: triangulation

The White and Barents Seas (Map: Chris Johnson)

in Oneshskaya Bay, topographic survey of the Kara Sea littoral (in a scale of 250 *sazhen*:1 inch; 1 *sazhen* = 2.13 metres); soundings by ship's boat near the Solovetskiye Islands, ship soundings from the vessels *Leitenant Ovtsyn* and *Leitenant Skuratov*, and the establishment of navigational markers.[10]

The fixing of navigational markers during those years had great value for shipping, especially in straits and at river mouths where they debouch into the sea. Thus in hydrographic activities during the 1904 campaign at the mouth of the Severnaya Dvina, this aspect of the work received much attention. It was carried out from the transport vessel *Bakan* under the command of Lieutenant-Colonel Lemiakov. In all, twenty-four leading markers, following the system devised by Major-General Vasiliev, were fixed at the river's mouth.[11]

The work of the Separate Survey of the White Sea continued over the years. During the summer months of 1907 the survey achieved the following: 76 linear *versts* (1 *verst* = 3,500′), and 132 square *versty*, surveyed; 2,376 point and 2,030 height determinations made; 1,414 linear *versts* sounded from ship, with 9,648 depth soundings and 1,580 determinations of points; 1,891 linear *versts* sounded from ship's boat, with 47,272 soundings and 2,687 determinations of points.[12]

In 1912 it was decided to undertake a systematic survey of the White Sea. The hydrographer-geodesist N.N. Matusevich was named to lead the work. The survey was interrupted by the First World War, but following the revolution, Matushevich continued this work as head of the Hydrographic Expedition of the White Sea. In specific terms, the expedition continued triangulation along the "Summer Shore" (i.e. the western shore), the topographic surveying of the mouths of the Dvina, the trawling of approaches to the Kandalaksha Bay, and other such activities. But the work was again interrupted, this time by the Civil War in 1918–20. At the close of the war, the activity of the Hydrographic Expedition of the White Sea was renewed in the summer of 1921, although with great difficulty. The triangulation of Kandalaksha Bay was carried out to the first-class level,[13] further work was done at the mouth of the Severnaya Dvina, channels were dredged, and so on.

Amongst other major difficulties facing the expedition was the issue of qualified personnel: there was a shortage of specialist-hydrographers. This was caused by human losses during the two wars, by the destruction of the nation's economy, and by famine and epidemics. Additionally, some of the skilled officers had by this time emigrated, while other specialists had been repressed by the Soviet state and were in prison camps. Hence the chief of the White Sea Expedition, in his report of 16 June 1921 to the officer commanding the Naval Forces of the Republic, requested that specialists found guilty of counter-revolutionary activity by revolutionary tribunals be allowed to serve out their sentences working in their areas of specialization, that is, performing services in the expeditions. This request specifically named the following individuals: the hydrographer V.V. Kostromitinov, sentenced to five years and at that time in the forced-labour camp of Kholmogory; the officer E.A. Stal'fel', a military topographer, in the same camp; and the hydrographer V.N. Andrianov, sentenced to five years' forced labour in Novo-Peskovsk camp; as well as others.[14]

Also, at that time and in part as a result of First World War activity, the safety of shipping in the White and Barents Seas was far from satisfactory. Between 1914 and 1918 the hydrographic service had recorded 101 shipwrecks involving ships of various nationalities (British, Russian, Norwegian, and others). Of these, 26 had run aground, 24 had been blown up by mines, and 25 had been sunk by German submarines.[15]

One should also stress that the logistical base for ensuring shipping safety was markedly weak. For the whole length of the White Sea littoral (fourteen hundred miles), the navigational safety network consisted of only the following elements: fifteen lighthouses, one floating light, twenty-four warning lights, thirty-five leading lights, and seven fog signal stations. Thus there was only one light for every thirty-five miles of shoreline, and only one fog station every two hundred miles.[16]

It is quite evident that these difficulties complicated even further the extremely burdensome conditions under which the hydrographic service of Russia's northern seas conducted its work. At the same time, life itself stubbornly demanded ever more from all areas of this difficult service. After the end of war, the scale of international trade began to increase, and the economy began to recover. This change, in its turn, increased the volume of shipping in the region. The increasing dynamics of Archangel's port activity can be seen from table 7.2.

Table 7.2

| *Year* | *Russian ships* | *Foreign ships* | *Total* |
|---|---|---|---|
| 1921 | 124 | 44 | 168 |
| 1922 | 265 | 90 | 355 |
| 1923 | 340 | 202 | 542 |

SOURCE: RGAMVF, Fonds r-457, Opis 1, Delo 17, 28.

As a result of these pressures, the Separate Northern Hydrographic Unit was created in 1923, again headed by Matusevich. Topographical surveys were continued through 1924–25, as well as soundings taken, in the Oneshskaya and Kandalaksha Bays. Work commenced on the study of the currents at the mouth of the White Sea, using Mitchells Floats and Ekman equipment, from the hydrographic ships *Kupava*, *Murman*, and *Metel'*. On the basis of this work, the *Atlas techenii v Gorle Belogo Moria* (The Atlas of Currents in the Mouth of the White Sea) was later published.

In 1925 the Separate Northern Hydrographic Unit was reorganized into the Northern Hydrographic Expedition. In 1926–28 it did virtually no work in the White Sea, as a result of a lack of funding. Only in 1929 did research commence again, and it increased in scale up to the outbreak of the Second World War. For example, during the 1932 season the following results were achieved: 175 linear kilometres of shoreline triangulated in Kandalksha Bay, 1,400 linear kilometres of ship soundings in the Severnoye Koshki Islands, and the erection of markers over 125 linear kilometres.[17] The years 1929–32 saw the following work accomplished: triangulation to the second and third levels, topographic surveys, and soundings in the delta of the Dvina, at the mouth of the White Sea, in the Mezenskaya and Kandalaksha Bays, and on the Karelsk, Terskiy, Letniy, and Zimniy shores.[18]

Thus, by the end of the 1930s, a new geodesic foundation had been established for virtually the whole shoreline of the White Sea, as well as a systematic descriptive survey of the Terskiy and Karelsk shores and the Kandalaksha, Dvinsk, and Mezenskaya Bays. Also, soundings of the open sea had been completed. From 1932 on, the White Sea Expedition was headed by the engineer-hydrographer B.I. Shamshur, while from 1934 D.V. Sishkov was in charge.

A few words need to be said about the hydrographic work carried out after the creation in 1933 of the Northern Naval

Fleet, which from the very beginning was based in the White Sea. It is known that ships of the Baltic Fleet constituted the foundation of the Northern Fleet. In large part, they reached the north by way of the White Sea–Baltic Canal. To survey this route in preparation for the passage of these ships, a "Separate Hydrographic Party" was organized under V.V. Kopasov. Major dredging of the whole route was undertaken – the Neva and Svir' Rivers, as well as the White Sea Canal. By July 1933 the route for warships travelling from the Baltic to the White Sea had been prepared, as well as the required bases and stopping points.

By about 1900, hydrographic knowledge about the shores of the Barents Sea had proved unsatisfactory and did not meet the growing needs of shipping in this region. Hence, in 1905 the Separate Survey of the Murmanskiy Coast was created, led by A.M. Bakhteev. Working from one small ship, nine officers commenced preparation of a new survey. A network of triangulations served as the basis of this survey. The officers completed a detailed plane-table survey of the shore line, took coastal and sea soundings, determined magnetic variation, and the like.[19] Employing these methods, between 1905 and 1914 they completed the study of the Murmanskiy coast from Kil'din Island to the Norwegian border.

For example, based on the work done in 1907, I. Bakhteev, chief of the Separate Survey of the Murmanskiy Coast, presented the Chief Hydrographic Administration with fifteen charts and plans. Plan 3 of the report charts covered the southern portion of Kola Bay, drawn to a scale of 250 *sazhen*:1 inch; plan 13 was a survey of the strait behind Bolshoi Olenii Island at the same scale. At the same time, Bakhteev deposited with the administration twenty "reporting" and "field" journals.[20]

The scale of work on the Murmanskiy shore increased with every passing year. By 1910 the Murmanskiy survey employed nine officers and ninety-three workers. Over the summer months of that year they achieved the following results: 200 linear and 84 square *versts* of shore surveyed, 450 linear and 66 square *versts* sounded from a ship's launch, 375 linear and 317 square *versts* of soundings from ship, six signals, forty-seven markers, 699 directional indicators set up, and other work done. The beginning of the First World War brought this activity to a stop.[21]

The Civil War and financial difficulties delayed the recommencement of the Murmanskiy survey until 1924, at which time the Murmansk Division was created, headed by the hydrographer-geodesist V.S. Lukianov. At this time, for the whole of the Russian Murmanskiy coast (650 miles) there existed only a very weak system for ensuring shipping safety. It consisted of the following: twenty-one lighthouses and warning lights, four leading lights, and two fog signal stations. Thus there was only one light beacon for every 30 miles of shoreline, and one fog station for every 325 miles of coast.[22]

Work now began to the east of Cape Kanin Nos, on the eastern Murmanskiy shore. Some aspects of the work completed in 1927 are as follows. In terms of triangulation, the following results were attained: nine signals established, twelve points studied, eleven points fixed, and eight centres established. A plane-table survey was conducted over a stretch of 100 kilometres of shoreline at a scale of 1:25,000. Launch soundings were done over 695 linear kilometres of shoreline, which included the taking of 12,725 soundings and the fixing of 1,968 points. Finally, ship soundings were completed along 433 miles, which included 1,417 soundings and the establishment of 651 points.[23] The scale of the work increased from year to year, and in 1929 the whole of the Murmanskiy coast up to Cape Svyatoy Nos had been triangulated to the second-class level. Topographic surveying had reached Zolotaia Inlet, while launch soundings had been performed to Baryshikhi.

During the 1930s, survey work along these lines continued further eastward. By 1934 the systematic survey of the Eastern Murmanskiy was completed and, combined with the survey of the Western Murmanskiy, finished even before the First World War, constituted a complete whole. Amongst those who led the Murmanskiy work, conducted under difficult conditions, one should note the roles of the hydrographer-geodesist I.D. Zhongolovich, of M.F. Ershov and G.A. Migalkin, and of V.P. Aleksandrov, the commander of the hydrographic ship *Kupava*.

Towards the end of the 1930s the hydrographers concentrated their efforts on the Murmanskiy coast on detailed work for large-scale charts. In addition, much was done to produce high-quality cartographic material for the needs of the Northern Naval Fleet.

By the beginning of the twentieth century, knowledge about the islands of the Novaya Zemlya group in no wise satisfied the needs of navigation. Unfortunately, given the financial difficulties, hydrographic work here proceeded at a far from satisfactory pace and in a fragmentary manner. For example, in 1901 a small expedition under A.I. Vil'kitskii undertook a survey of the east coast of Novaya Zemlya, but only from Matochkin Shar and Medvezhiy Bay northwards. Later, in 1910, another expedition under Lieutenant G.Ia. Sedov effected a simplified plane-table survey and occasional soundings in Krestovaya Inlet. In 1912–13 this same group did a route survey from the Pankrat'yeva Peninsula to Cape Zhelaniya.[24]

After the First World War, hydrographic investigation recommenced in 1918–19 in the region of Novaya Zemlya. However, all the results of this work were lost because of the Civil War, as it unfolded in North Russia. Once the war was over, the need to develop commerce and to search out mineral resources served as the rationale behind hydrographic work on Novaya Zemlya. In 1921, under the leadership of N.V. Roze, an expedition began preparing for the construction of a radio station on Cape Matochkin Shar. At the same time, a marine survey of the east coast of Novaya Zemlya, from Blagopoluchiya Bay to Cape Zhelaniya, was completed. In 1923 a radio station and a geophysical laboratory were built at the eastern entrance to Matochkin Shar. This facility established the practice of serving the Northern Sea Route by providing information to ships on ice and navigational conditions.

Only in 1924 was it possible to begin work on a systematic survey of Novaya Zemlya under the general leadership of N.N. Matusevich. The following year a plane-table survey was completed from Matochkin Shar northwards to Volchikhi Inlet and from Cape Britvin to the Khramtsov Peninsula. In 1926 a plane-table survey was conducted in the area from Volchikhi Inlet to Sukhoy Nos and Safronova Inlet, the shoreline of Mollera Inlet from Cape Britnev northwards to Bezymyannaya Inlet, and from Cape Karelskii southwards to Severnyy Gusinyy Nos. Soundings were generally taken along all these coasts from a launch,[25] and the shores were surveyed. During the 1929 season it proved possible to survey 160 kilometres of shoreline at a scale of 1:50,000.[26]

Broadly speaking, hydrographic work on Novaya Zemlya was conducted irregularly, with large interruptions in-between. In 1922 no work was performed here, as was the case in 1927–28, while work begun in 1931 also had to be interrupted. In 1932

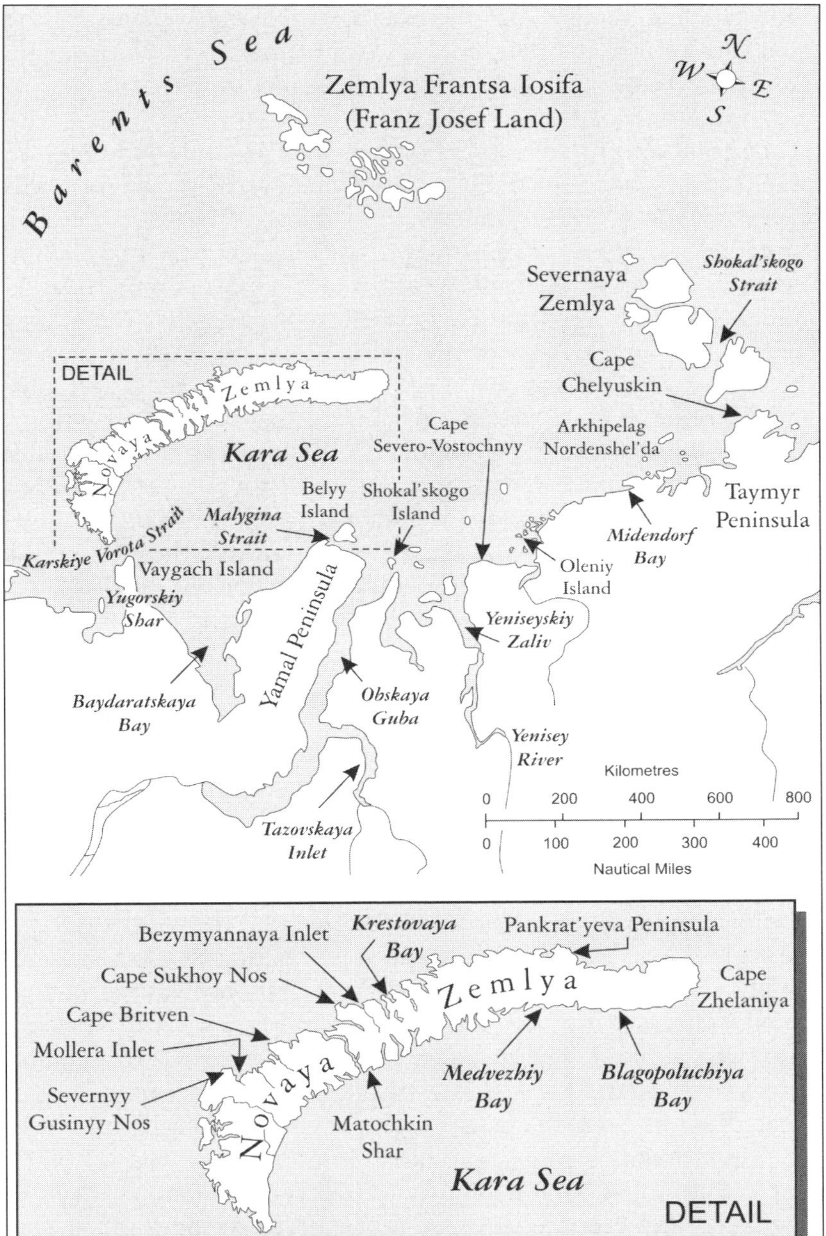


The Kara Sea: Novaya Zemlya to Severnaya Zemlya
(Map: Chris Johnson)

hydrographic studies of the island of Novaya Zemlya were renewed under the engineer-hydrographer B.I. Shamshura. Twenty astronomical points were determined, which served as the foundation of future work.[27]

The product of ten years' efforts, from 1923 to 1933, was the preparation of cartographic material that satisfied current needs regarding the southern and western coasts of Novaya Zemlya. Nonetheless, the eastern littoral of Novaya Zemlya was not completely charted; detailed survey information existed only for individual bays and anchorages. It is worth noting that the survey of 1923–33 of Novaya Zemlya was based on triangulation to the third class and was accompanied by hydrographic, meteorological, gravitational, astronomical, and magnetic observations.

At the start of the twentieth century, hydrographic knowledge and the provision of navigational support for the Northern Sea Route did not at all answer the navigational needs of the time. There existed only a few navigational charts of the seas of the Arctic Ocean. Moreover, the coastlines as depicted on them had been determined on the basis of knowledge garnered from land routes and from some marine surveys; only an insignificant portion of the chart information was based on at least partial measurement by instruments and on astronomical points. Depths were marked very infrequently, based only on incidental soundings. No pilots had been completed for the Northern Sea Route; hence sailors had no possibility of reliably determining their location from coastal landmarks.

To resolve these urgent problems connected with the hydrographic study of the Northern Sea Route, the Arctic Ocean Expedition was established in 1898. It was first headed by Lieutenant Colonel A.I. Vil'kitskii of the Corps of Naval Navigational Officers. Vil'kitskii was replaced in 1902 by A.I. Varneko, who was himself succeeded in 1903–05 by F.K. Drizhenko. During the years 1898–1910 this expedition accomplished a great deal, in both the scope and the importance of its work. Beginning from fixed astronomical points, it surveyed Obskaya Guba and Yeniseyskiy Zaliv; it studied the channels in Pechorskaya Bay and prepared a plane-table survey of the southeastern shore; it sounded the straits of the Karskiye Vorota and Yugorskiy Shar and completed a plane-table survey of the Yugorskiy Shar as well; it effected a marine survey of the western and eastern shores of Vaygach Island and soundings and a survey of the northern portion of Baydaratskaya Bay; and finally, it compiled a marine survey of the western and northern littorals of Yamal Peninsula.[28]

In 1904 the Arctic Ocean Expedition consisted of seven officers, forty-seven sailors, and nine other employees working from three ships – *Pakhtusov*, *Leitenant Ovtsyn*, and *Leitenant Skuratov*. F.K. Drizhenko led the expedition at that time. Hydrographic studies were conducted in the Karskiye Vorota and along the northern shore of Vaygach Island, while a marine survey was conpleted along the western shore of the Yamal Peninsula.[29]

It is worthy of note that in 1900–02 the Russian Polar Expedition was also under way, led by E.V. Tol. It had been organized by the St Petersburg Academy of Sciences, thanks in large part to the efforts of Grand Duke Konstantin Konstantinovich, its president. This expedition, in particular, managed to survey a portion of the west coast of Taymyr Peninsula, the Arkhipelag Nordenshel'da, and one of the inlets of Kotel'nyy Island. One of the members of this expedition was the famous hydrologist and magneticist A.V. Kolchak.

Somewhere around 1907, within the Chief Hydrographic Administration, the idea was born of organizing a large-scale expedition for the purpose of studying the Arctic Ocean. Its chief initiators were A.I. Vil'kitskii, head of the administration, the renowned hydrographers F.A. Matison and A.V. Kolchak, and several others. For the purposes of this objective, the Naval Ministry managed to obtain funding for the construction of two icebreakers. In February 1908 the ministry concluded a contract with the Neva Shipbuilding Yards for the construction of the vessels, at a cost of 1,320,000 rubles, with completion specified for "not later than two weeks after the opening of navigation at Kronstadt in 1909."[30] To jump ahead somewhat, the shipbuilders did not manage to meet the deadline for completion of the icebreakers. Final trials of the vessels, which were named *Taymyr* and *Vaygach*, were completed only in the autumn of 1909. The following April these ships, by order of the Naval Ministry, "were transferred from the Baltic Fleet to the Siberian Flotilla" and departed for their base, Vladivostok.[31] *Taymyr* and *Vaygach* undertook their first expeditionary work in hydrographic surveying of the Bering Strait in September–October 1910.

During the years 1910–15 the Arctic Ocean Expedition continued its work, which was now done with the specially constructed *Taymyr* and *Vaygach*. From their Vladivostok base, these icebreakers sailed annually to carry out hydrographic studies from Dezhneva Cape westwards. Over these years the expedition prepared a marine survey of part of the Siberian coast from Bering Strait to Cape Chelyuskin and portions of the Novosibirskiye Islands, discovered Severnaya Zemlya in 1913, and surveyed the eastern and southern shores of this archipelago. These vessels also discovered three hitherto unknown islands.

In his "most humble report" to Czar Nicholas II, the naval minister of the time, I.K. Gregorovich, informed the czar of the discovery of the three new islands. Somewhat later he proposed to the czar to name the islands as follows: the largest of the group after Nicholas II himself; the second in honour of the czarevich, Aleksei; and the third after A.V. Kolchak, in honour of his heroic attempt by whale boat to save E. Tol in 1903. Czar Nicholas agreed to the first two recommendations, but decreed that the third island be named after General A.I. Vil'kitskii.[32]

Right up to the time of the Revolution of 1917, along the whole of this massive stretch of Siberian littoral, a system of navigational supports was almost non-existent. For example, for all the coasts of the Kara Sea, there were only fifteen navigational markers. The shores of the Laptev, East Siberian, and Chukchi Seas had, in total, only ten navigational markers, with only approximate coordinates. A mere four radio stations serving shipping were located in the western part of the Kara Sea.[33]

Even under the most difficult of conditions during the Civil War in 1918–19, the Arctic Ocean Expedition never ceased its work. Specifically, it continued surveys along the west coast of Novaya Zemlya, as well as fixing channel markers in Yugorskiy Shar and exploring the mouth of the Yenisey River.[34] At the close of the Civil War, towards the end of 1920, the Separate Ob-Yenisey Hydrographic Division was detached from the expedition and commenced work under the hydrographer K.K. Neupokoev. This division specifically assisted the so-called Kara Grain Expeditions, whose purpose was to ship foodstuffs from Siberia to European Russia.

On the whole, most of the work done was concentrated in the "southern" straits, which joined the Barents and Kara Seas. For

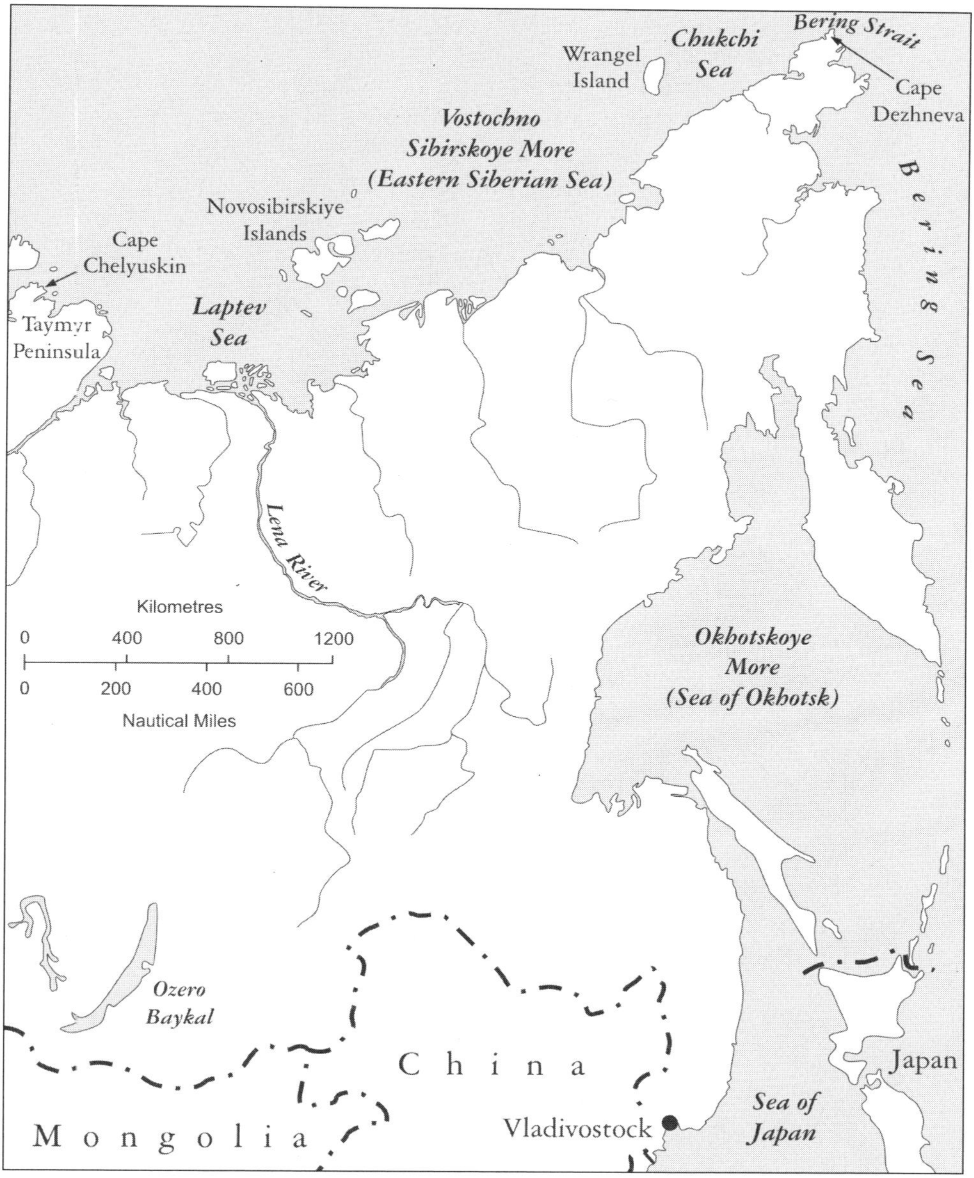


Cape Dezhneva, Bering Strait, to Cape Chelyuskin
(Map: Chris Johnson)

this specific purpose, in 1920 the Karskiye Vorota Division was established, which had at its disposal the ships *Taymyr* and *Pakhtusov*, the sail-and-motor vessel *Bednota* (formerly *Dmitrii Solunskii*), and the motor boat *Arktur.* Insofar as material support was concerned, the unit was extremely poorly supplied, and it had great difficulty securing foodstuffs. Russia was then under the rule of "war communism," when the use of money was banned. Food was supplied in the form of rations, which were so small that they could barely sustain the semi-starving hydrographers. Hunting literally saved the expedition from starvation. There was, in particular, a great abundance of various game birds in the north. In one of his reports, the commander of *Taymyr* recorded that, at one of the ship's stopping places on the Kara Sea coast, the ship's crew went ashore and, in the space of one hour, "clubbed to death 150 geese."[35]

In 1922 the Arctic Ocean Expedition was disbanded and replaced by the Administration of Navigational Safety in the Kara Sea and in Siberian River Mouths (UBEKOSIBIR). This administration was headed, at various times, by the following specialists: K.K. Neupokoev (1922–23), N.F. Timofeevskii (1926–29), and I.V. Andreev-Dolgov (1930–33). In the period from 1922 to 1933 the hydrographers of UBEKOSIBIR accomplished a great deal; the following achievements are particularly worthy of note: the survey of the Obskaya Guba, with the Belyy and Shokal'skogo Islands, Malygina Strait, and Tazovskaya Inlet; the survey of the coasts of the Taymyr Peninsula, from Cape Dvukh Medvedei to Midendorf Bay and the Severo-Vostochnyy Islands; and the survey of Yeniseyskiy Zaliv, including the islands of Vil'kitskogo, Neupokoyeva, Sibiryakova, Oleniy, and others.[36] In sum, these hydrographers surveyed more than five thousand kilometres of continental coastline and islands, basing their work on astronomical observations and local geodesic networks. They also took soundings over eighteen thousand kilometres. At the same time, they conducted magnetic, gravimetric, hydrological, and meteorological observations.

During these years of systematic hydrographic study in the Ob-Yenisey region, a number of specific expeditions were also organized, which were very significant for the development of the Northern Sea Route. In 1924, for example, an expedition under the leadership of the hydrographer-geodesist B.V. Davydov was organized to conduct scientific investigations on Wrangel Island. In Soviet historiography one comes upon the notion that the Soviet government at that time had designs on Canadian hunters, then "poaching" on Wrangel Island.[37] Thus B.V. Davydov's expedition was also enjoined to remove these Canadian citizens from the island.

The scale of hydrographic activity in the Arctic Ocean gradually increased. During the 1927 season, for example, the following work was completed: in Obskaya Guba a plane-table survey of 145 miles of shoreline (at a scale of 1:84,000) and ship soundings of 1,800 square miles, and in Yeniseyskiy Zaliv triangulation of over 200 kilometres of shoreline, the establishment of twenty markers, and the laying out of three triangles, as well as astronomical and magnetic observations at sixteen points.[38]

In 1932 an expedition under A.M. Lavrov on the *Taymyr* was organized for the northeastern part of the Kara Sea and the straits that link this body of water with the Laptev Sea. The chief objectives of the study were the Shokal'skogo Strait, the northwestern coast of the Taymyr Peninsula, and the Nordenshel'da Archipelago. This expedition discovered eight islands, determined seven astronomical and five magnetic points, conducted studies of the gyro compass and of echo-sounding (asdic) under conditions of Arctic navigation, studied ship-handling in ice, and the like.

In looking at records of the Northern Hydrographic Expedition of 1932, we discovered mention of a unique historical event. The expedition's report of that year specifically noted that there was no need to conduct any triangulation of the coastline of Vaygach Island,[39] because another hydrographic service had already been at work on the island for several months. This was the Vaygach Hydrographic Expedition of the Labour Camp Administration of the OGPU. The giant machinery of repression initiated by the totalitarian regime had already been set into full motion, and the OGPU itself was exploiting new territory in the far north for the construction of concentration camps intended to utilize the labour of hundreds of thousands of convict-slaves for the full and final victory of socialism "in one country."

In December 1932 the Soviet government established the Chief Administration of the Northern Sea Route in Moscow (GLAVSEVMORPUT'), followed in the spring of 1933 by the creation of the Hydrographic Administration of GLAVSEVMORPUT', headquartered in Leningrad. These organizations were charged with the task of developing a sea route and ensuring the safety of shipping from the Novaya Zemlya Straits to the Bering Strait and equipping the route with modern navigational devices. Between 1933 and 1941 the hydrographers of GLAVSEVMORPUT' conducted more than eighty expeditions, and by the beginning of the Second World War the systematic survey of the Arctic seas was, in the main, completed.

How were the results of these surveys used? In pre-revolutionary Russia the best result in this area was achieved in 1913, when ten new charts were published. In the Soviet period the publication of new charts can be summarized as follows: 17 charts in 1929, 78 charts in 1933, and 168 charts in 1937. In addition, while in 1913 the total print run of charts was ten thousand copies, in 1937 it reached 1 million copies. One should also mention, however, that in the USSR right up to the end of the 1930s, foreign-produced charts were in wide use.[40]

Towards the end of the 1930s the hydrographic service of the USSR printed various pilots, such as *Information for Seafarers*, descriptions of lights and markers, *Tide Annuals*, numerous specialized tables, and other such aids. The volume of these publications in the 1920s and 1930s was as follows: in 1925 a total of 130 printed pages were produced, in 1929 the figure was 134 printed pages, in 1933 it was 345 printed pages, and in 1937 it reached 599 printed pages. (A printed page equals 22 typed pages).[41]

In conclusion, there is sufficient evidence to suggest that in Russia and the USSR, over the period under study here (1900–40), hydrographic surveying of the northern seas at the government level was conducted intensively and with positive results. This work was shaped by the needs of navigation in the service of the national economy, by the needs of the Navy, and in the interest of studying and developing the vast territorial expanses and natural resources of Siberia and the north. The results of these studies have come to serve as a sound, reliable foundation for the development of hydrographic knowledge of the northern seas in the post-war period.

# 8

# *Wartime German Hydrography in Canadian Waters*

MICHAEL L. HADLEY

The earliest German engagements in what we now recognize as Canadian waters and their contiguous zones began either as experiments in cartography or as participation by individual Germans in explorations conducted by other nations. These scientific and technical speculations during the sixteenth through eighteenth centuries were but the prelude to explorations triggered by the German government's own strategic and tactical interests in an age of major-power war.

Martin Waldseemüller's world map of 1506, for example, provided the first "German" glimpse of what might be imaginatively called a Pacific Ocean. Based on information available at the time, his map has been described as an inspired guess.[1] It served as a source for many delineations. Notable among these is the 1520 map of Nürnberg astronomer and mathematician Johannes Schöner. It shows an imaginary continent (North America) amidst great waters, thus depicting an Oceanus Occidentalis and an Orientalis Oceanus – an Atlantic and a Pacific Ocean. As Waldseemüller had done, Schöner represented the west coast as a straight line. Given the increasing significance of the search for "Cathay" and a Northwest Passage, the eighteenth century paid particular attention to the Asia-Pacific region. Thus Nürnberg cartographer Johann Baptist Homann incorporated many ideas of the day into his commercially produced maps and atlases, including Hermann Moll's map of 1719. Among other speculations, Homann's map of 1731 revealed a fanciful west coast of North America. Actual experience of the area was left to Gerhard Müller, a participant in Vitus Bering's Second Kamchatka Expedition of 1741. Müller actually journeyed across the top of the Pacific and published a map in 1758 that presented the most definitive picture at the time. He is considered to have been perhaps the most knowledgeable person about the geography of the area. As a recent scholar explains, "his concept of the North Pacific remained the authority until James Cook's third voyage" (1776–79), which aimed at finding a passage north over the top of North America from the Pacific.[2]

Significantly, Cook had been perplexed during the voyage by the error-ridden map published by another German, Jacob von Stählin. Map-making in the eighteenth century seems to have had a ready market. Inspired, for instance, by widespread interest in Cook's third voyage and nourished by information supplied by crew members, cartographers produced maps well before the appearance of Cook's own account. Key among these contributions was the map of Nürnberg cartographer Tobie Conrad Lotter. His 1781 version of the North Pacific showed, by his own account, "the coasts which bound it on both sides according to the most recent discoveries made by Spanish, Russian and English up to 1780."[3] But there would be no German ventures into Canadian waters for more than two hundred years.

German explorations of Canadian waters in the twentieth century had a dual motivation: the preparation for war and the actual conduct of maritime operations against Canadian and Allied forces. From 1904 to 1945 German mariners and submariners experienced and reported on a variety of challenging conditions, ranging from survey on the BC coast to Arctic intrusions to under-ice navigation in the St Lawrence, cable-cutting on the Grand Banks, covert operations in the Baie des Chaleurs, transit of the Strait of Belle Isle, and penetration of Wabana anchorage, Newfoundland. (Contrary to Canadian folk wisdom, no German war vessels operated off the west coast of Canada during the war; nor did Germany build any U-boat pens in BC coastal waters.) Implementation of the emerging intelligence picture was limited solely by the strategic and tactical goals of the day.[4] Gathered in the heat of battle, hydrographic information was disseminated in briefings and navigation packs. Germany's war navy, the Kriegsmarine, was defeated by the Allies in May 1945, thus preventing any immediate post-war evaluation of hydrographic experience. Thus hydrographic intelligence remained archived in disparate and dispersed files, and lingered in the memories of veterans. Ten years later, in 1955, its successor, the Federal German Navy (Bundesmarine) was established and began to examine its roots. As will be seen, not until 1987 did the German navy of the day inquire in open forum into the "German experience" of so-called shallow water operations off Canadian shores.

The story begins with Imperial Germany's concerns about a potential war with the United States. As early as 1903, German survey ships were exploring foreign coasts in search of possible locations for covert "support anchorages" – *Unterstūtzunsplätze* – where German cruisers could replenish with coal from friendly ships and prepare for combat. In April 1904 the Kaiser himself approved the exploratory voyage of the small survey ship SMS *Falke* "to all the more important coastal points [in the United States] including the west coast of Canada."[5] In the event of war, so German naval leaders thought, German forces would invade the United States from Canada and engage in a territorial campaign. During 1905–06 the gunboat SMS *Panther* explored the west and east coasts of the United States and Canada. Alongside in Seattle on 16 September 1905, her skipper wrote a report on her scouting forays along the BC and Alaskan coasts. He claimed to have discovered on the BC coast "a large number of hiding places and protected harbours ... of special importance in the event of war against the United States and against Canada."[6] By November 1906 SMS *Panther* was charting and photographing in Saint John, New Brunswick.

In a secret, handwritten document the chief of staff of the German Admiralty ordered the commanding officer of the cruiser SMS *Bremen* in June 1907 to scout the harbours of Halifax and Quebec City. "All available war charts of the area," the chief complained, were "sketchy and of older origin." He therefore required SMS *Bremen* to update them. Scouting operations were thorough for the day. Starting from Sambro Island, the Bremen's intelligence-gathering naval units worked their way from seaward to the inner harbour of Halifax, even examining the Eastern

Passage. (Suggestions by locals that Royal Navy cruisers regularly navigated the Eastern Passage at night led to a serious German survey the following year.) *Bremen*'s skipper sent surveying parties ashore in both Halifax and Quebec City. His report to the Kaiser, written on 10 October 1907, revealed intimate knowledge of Canadian defences. (These defences were, of course, rather lean.) The Kaiser expressed satisfaction with the tour of SMS *Bremen*, particularly with her visit in September 1907 to Montreal, which, in his words, had "helped secure the good repute of the German navy abroad."[7] Other German warships followed: for example, the cruiser SMS *Freya* in August 1908.

Precisely how much information from these surveys actually filtered into operational navigation packs prior to the outbreak of the First World War is difficult to ascertain. Suffice it to say, however, that by July 1916 Germany's submarine freighter *U-Deutschland* was crossing the Atlantic from Wilhelmshaven to Chesapeake Bay in twenty-four days, thus establishing a submarine supply line that undercut British cruiser deployments. She operated comfortably in very shallow coastal waters. (Her sister ship *U-Bremen* was lost in August 1916 on her maiden voyage.)[8] The combat submarine *U-53* followed in October 1916, sinking ships in shallow waters off New York with impunity.[9] The story in Canadian waters was similar. Virtually unopposed, German submarine cruisers armed with 5.9-inch cannon operated with impunity, even off the entrance to Halifax Harbour.[10] They laid mines in shallow waters and cut transatlantic cables. Both operations required precise navigation. Drawing on adequate charts, Pilots, and military intelligence sources, they experienced no serious difficulties in coastal waters.

Cable-cutting is a special case that highlights the Germans' detailed knowledge of the shallow waters of the continental shelf. The former submarine freighter *U-Deutschland*, now converted into a powerful U-cruiser and renamed *U-155*, departed Wilhelmshaven on 11 August 1918 with specific orders. She was to lay mines in Canadian inshore waters and to "cut the cable northwest of Sable Island," as her war diary notes. (The desolate, twenty-mile-long crescent of sand lies some one hundred and fifty miles east-southeast of Halifax.) As a detailed study explains:

> *U-155* worked her way along the route from Halifax to the northeastern side of Sable Island in order to locate the transatlantic cable. German

> intelligence had rejected other alternatives: Heart's Content, Newfoundland, had little strategic value, despite being "technically the simplest to cut"; and the St. Pierre cable was "too silted up after 40 years on the bottom." … Now working toward Sable Island, [the skipper] sought a spot where the cable lay in a soft seabed that would not damage her experimental and apparently clumsy cutting gear. Two attempts in rocky ground snapped *U-155's* rig; but on 19 September [he] found favourable sandy ground. His log records his "surface run on cutting course 130 [degrees] in order to begin cutting the innermost of the six northwesterly Sable Island cables in 90 meters of water." Emergency dives when exposed to "steamers, destroyers and darkened ore-carriers" interrupted his work, until the tension indicator on his rig hovered in a range of "1.8–2.6 tons [and] indicated a sinewy object had been gripped and cut."[11]

Though the log contains few details about the rig, it would doubtless have been similar to that recorded in the log of the U-cruiser *U-151*, which had cut cable close in to New York in May 1918, in depths of twenty-five fathoms. The rig consisted of a derrick, winches, greased flexible steel wire rope, and pressure cutters capable of biting through the cable's thick protective skin and armoured strands.[12] Significantly, Canadians and Americans repaired the cut cables within a matter of days.

German records covering the interwar years offer no clear evidence of hydrographic surveys. Certainly, between the scuttling of the interned German fleet in Scapa Flow in 1919 and the Anglo-German Naval Agreement of 1935, the German navy had few resources. The secret development of submarine technology in the early thirties reminds us, however, of the extent of the navy's covert resources during these years. Nonetheless, we have no hard evidence of any special emphasis on hydrographic intelligence gathering – neither by German merchant shipping operating in Canadian waters nor, indeed, through links with the Imperial Japanese Navy. Yet commercial shipping would have gathered information in the normal course of conducting its maritime business. It is striking that during the Second World War, German naval forces had recourse to detailed charts and Pilots. Veterans still recall putting to sea with reliable navigation packs, though often the scale of the charts – large area, small scale – made inshore operations difficult.

Germany published its own German Admiralty charts (*Deutsche Admiralitätskarten*), known as *Segelkarten*, or "sailing charts."

(The non-technical reader may wish to skip this paragraph.) The large-area chart 442, covering Halifax and contiguous waters including the Virgin Islands (*Halifax bis Virgin Islands*), first appeared in 1912 on a scale of 1:2,500,000; it was reissued in 1939. The northern portion of Newfoundland, chart 444, entitled *Belle Isle–Neufundland*, was published in 1913 on a scale of 1:600,000. It corresponded to the areas covered by British Admiralty charts 3335 and 779. The southern portion – 445, *Neufundland–Cabot Straße*, also listed as *Neufundland, Südlicher Teil* – was drawn to a scale of 1:600,000; it was first issued in 1912. Nova Scotia waters, covered by 447: *Neuschottland*, appeared in 1912 on a scale of 1:600,000 and was reissued in 1934. Chart 446 for the Gulf of St Lawrence (*St. Lorenz-Golf*), on a scale of 1:1,000,000 and roughly corresponding to British Admiralty charts of the day 1134, 1623, and 2034, first appeared in 1913; it was reissued in 1935. German chart 448, *Golf von Maine-Fundy Bucht*, on a scale of 1:625,000, first appeared in 1913, roughly equivalent to British Admiralty charts 352; it was reissued in 1935. The St Lawrence River itself, an area corresponding to British Admiralty charts 307 and 312, was covered by German chart 446: *St. Lorenz-Strom*. All of these charts were in use throughout the war. Included in a lengthy and comprehensive series of "Naval Service Regulations" (*Marinedienstvorschriften*), embracing everything from technical manuals to summaries of the "War experiences of German submarines, 1914–1918," were two of navigational significance: a submarine handbook of the east coast of Canada and an atlas of the same area.[13]

Remarkable for their clarity are five Pilots, or navigational handbooks, published by the Naval High Command during the Battle of the Atlantic. Two of these Pilots, published in 1941 and 1943, cover the east coast of the United States;[14] and three of them, published in 1942 and 1943, the east coast of Canada and the Gulf of St Lawrence. Significantly, the editions covering Canadian waters bore the titles "Submarine Navigation Handbook."[15] The first "Canadian" volume (published in August 1942) covered the northeast and southeast coast of Cape Breton Island, the southeast coast of Nova Scotia, and the Bay of Fundy; the second (printed in September 1942) dealt with Newfoundland and the Strait of Belle Isle. Significantly, these volumes appeared seven to eight months after the opening submarine salvos of "Operation Drumbeat."[16] This was Germany's first strategic advance on the Atlantic coast of North America. Eight submarines formed the first wave from 12 to 31 January (with five more providing its support wave from 8 January to 12 February); eight submarines formed the second wave from 21 January to 19 February, and thirteen submarines the third and final wave of this winter phase. U-boats in these waves sank ninety-seven merchant vessels, dislocated convoy operations, gained geopolitical advantage, and, of course, gathered vital navigational information for successive forays. The last of Germany's "Canadian" navigation handbooks, a "submarine Pilot" covering the Gulf of St Lawrence, was published in October 1943, over a year after the so-called Battle of the St Lawrence had begun. The "submarine atlas," covering the east coast of Canada and containing thirteen detailed charts of the Bay of Fundy, did not appear until August 1942, three months after *U-213* had operated in the bay and landed its spy at Melvin's Beach, near Saint John, New Brunswick. (Though the data was never used, the submarine atlas paid detailed attention to the Bay of Fundy as a target for mines.) When *U-518* landed a second spy on the beach near New Carlisle, Quebec, in November 1942, it had no recent German edition of the Pilots; thus the skipper might have resorted to a British edition as well as to German and British Admiralty charts based on a survey of 1913. Still, ample navigation lights led him along the Miscou banks on the southern shore of the bay before he cut across and placed his bow on the sloping shore with a rising tide.[17]

From January 1942 until April 1945 German submarines undertook a series of daring and complex operations in Canadian inshore waters. They sank merchant ships and warships, laid mines, operated under the ice of the Cabot Strait, advanced into the St Lawrence as far as Pointe-au-Père, penetrated Conception Bay and Wabana Anchorage, and established an automatic weather station in Labrador. They undertook special operations: landing spies in New Carlisle on the Baie des Chaleurs and at Saint John, New Brunswick, and two failed attempts at picking up escaped prisoners of war from Prince Edward Island and Maisonette Point on the Baie des Chaleurs. Many of these ventures were exploratory in nature. For example, the "Battle of the

St Lawrence" itself (the term was coined by the *Ottawa Journal* in 1942) was triggered by the curiosity of U-boat commanders, entering via either the Strait of Belle Isle or Cabot Strait, who simply wanted to know what was inside. These intrusions were essentially independent spinoffs from larger operations. In all cases, radio contact with U-boat Command in Kerneval, France, together with detailed War Diaries (*Kriegstagebüchen*) and oral debriefings on return to port, apprised Naval High Command of both tactical and navigational conditions.

Penetrating into the Gulf of St Lawrence, *U-553*, the first of the intruders, had by 9 May 1942 reached Anticosti Island. Despite mists and occasional fog, the submarine encountered no navigational difficulties at all. Indeed, its war diary reported that Canadian "lighthouses and radio beacons were functioning under peacetime conditions"; or again, "radio beacons functioned under peacetime conditions according to the current List of Lights and Radio Signals."[18] With slight exception, Canadian navigational aids remained available to navigators throughout the war. The following month, Grand Admiral Karl Dönitz directed *U-132* to scout a critical "choke point" of the St Lawrence, an area designated by a line running between Pointe-des-Monts on the north shore and Sainte-Felicité on the south. In the event, *U-132* would exceed this advice by reaching as far inland as Pointe-au-Père, some 172 miles from Quebec City. A series of modern Canadian hydrographic charts annotated to show the tracks of German submarines in east-coast waters and in the St Lawrence River and Gulf is deposited in the Canadian War Museum in Ottawa.[19]

German submariners would seem to have been the first to discover the tactical significance of temperature and density layers in the St Lawrence and off the Atlantic coast of Canada. Transiting the Strait of Belle Isle in late August and early September 1942, German submarines *U-517* and *U-165* found similar conditions. *U-517* reported unexpected density layers near Forteau Bay, which caused it to plunge to the sandy bottom when diving. Indeed, the skipper described being hunted by a Canadian vessel which, in the German's mind, "ought" to have detected him. Technically at least, the Canadian's asdic (as sonar was then known) had left the Germans in no doubt that they had been detected: the vessel circled the submerged U-boat repeatedly without dropping a single bomb, probing with her asdic's threateningly clear chirping and rasping tone, which, on close approach, crescendoed to the unnerving sound of fine gravel being cast along the submarine's hull. Likewise, the War Diary of *U-43* recorded, while in the St Lawrence River during October 1942, that "the considerable water stratification" protected the boat from attack even when the hydrophones could hear the attacker's propellers and when the submariners could hear the "pinging" of sonar. As the Germans reported, all these near misses bore the mark of "untrained escort vessels" which were having serious difficulties with sonar detection. Recently, Marc Milner has analyzed the situation and explained: "Asdic conditions in the Gulf of Saint Lawrence and in the mouth of the river were notoriously poor. It was impossible, for example, to obtain asdic contact on U-boats immediately following attacks, even though the general location of the submerged submarine was known."[20]

Meanwhile, Paul Hartwig's *U-517* entered Forteau Bay in the Strait of Belle Isle, which his navigation instructions described as a suitable anchorage for merchant vessels at the western end of the strait; during an emergency dive, he encountered unexpected density layers that sent him plummeting into the sandy bottom. Germans quickly discovered the gulf's special properties for distorting sonar. The allied warships in Canadian waters suspected as much. As HMS *Salisbury's* report noted after the rash of sinkings by Hartwig's *U-517*, "since most of our ships have been sunk westward of Gaspé Passage, the extremely poor [sonar] conditions must be known to the enemy who have often fired their torpedoes during the day and made good their escape."[21] Hartwig's War Diary reveals that he had frequent occasion to curse the density layers when attempting to dive or surface. He literally had to punch his way through them in a dynamic descent or ascent under power. With impunity, *U-513* attacked vessels in Wabana Anchorage in September 1942, while *U-69* explored Canadian inland territory: from Cabot Strait, past Baie Comeau, and then off the tidal flats of Manicouagan River (Battures de Manicouagan) to Matane. On 19 October 1942 *U-43* reached as far as Pointe-au-Père, the westernmost point on the river reached by U-boat at the time. Throughout all these operations, all submarines were required to send to their commander-in-chief regular weather reports that included the temperature of sea and air. Judging by the constant references to mist, fog, and ice, their Canadian patrol was with slight exception a grim business indeed.

These U-boat incursions of 1942 highlighted for Canadians the need to investigate the problem of sound propagation in east-coast waters, and they triggered Canadian initiatives in bathythermography – investigation of the changing temperature of the ocean in relation to depths. In the fall of 1943 the anti-submarine officer based in Gaspé undertook a rudimentary review of asdic operators' experiences in attempting to detect German submarines. As Milner has reported, "His entirely unscientific results told the navy what it already knew: asdic conditions inside the 100-fathom line were dreadful for all but a few months in spring and fall."[22] Here, as elsewhere, Milner has revealed Canadian naval suspicions about the value of the infant science of oceanography. In the words of the staff officer (anti-submarine) based in Halifax in December 1943, whom Milner cites for the previous passage, "no practical results are at present being aimed at and in addition I am extremely doubtful if any practical results can ever be achieved." While the National Research Council had been involved with the Royal Canadian Navy's (RCN) bathytherographic interests since 1942, it was not until May 1944 that "a new 'Atlantic Oceanographic Research Group' was established in St. Andrew's, NB, ... to begin serious work on sound propagation in Canadian waters."[23]

When *U-536* entered the Baie des Chaleurs in September 1943 in an attempt to pick up German prisoners of war who had escaped from Bowmanville, Ontario, he found his charts out of date and important landmarks missing. Yet when aborting the attempt because of the trap that the RCN had set, he escaped as close inshore as possible. The record suggests that the submarine actually skidded along the bottom near the Miscou flats in twenty metres of water, at one point even becoming entangled in a fisherman's nets. This seems to have been one of the rare operations when a commander complained of inadequate navigational charts. Of course, the year 1943 was the turning point in the Battle of the Atlantic. A confluence of superior Allied technologies was making it difficult for Germany to wage war at sea. In the spring of that year Grand Admiral Dönitz had actually recalled all his boats from combat. Only the deployment of what the Allies called the "Gnat" target-seeking torpedo

(*Zaunkönig*) and the snorkel gave Germany hopes of prevailing. Weather reporting was a crucial factor. The German Naval Weather Service (Marinewetterdienst, or MWD) had long been actively engaged in establishing and operating weather observation stations in the Arctic.[24] Recognizing the severe weather conditions prevailing in the western Atlantic, particularly around Newfoundland, the MWD decided to expand its Arctic chain of stations by positioning an automatic station in Labrador.

*U-537* set up its automatic weather station WFL 26 (WFL is an acronym for *Wetterfunkgerät-Land*, or land-based weather radio) in a very remote part of Labrador in October 1943. This unmanned station constituted a vital link in a chain of Arctic and Subartic stations based in Spitsbergen, Bear Island, Franz Josef Land, and Greenland. The choice of the actual site in Labrador seems to have been left to the skipper of *U-537* in consultation with his technical advisers. He wanted to be as far north as possible in order to escape detection. As his War Diary records, he made landfall at "Kap Shidley" (Cape Chidley) on the northwestern tip of Labrador after five days of dead-reckoning navigation in heavy snow. He then coasted southward along the poorly charted littoral, relying on his echo sounder in areas that, even today, abound in uncharted reefs and shoals. His charts had led him to suspect the skerry, shale coastline that greeted him. Still navigating by echo sounder, he eased his way southeastward between Home Island and what he called "Arayal" (actually Avayalik Islands). Rounding the southern tip of Hutton Peninsula, he finally anchored in "Attenaukjuke Bay" (now Martin Bay). Within twenty-four hours, the automatic weather station, code named "Kurt," was operating. This is a rare example of a covert operation that may have contributed hydrographic information to the German Hydrographic Institute, which in turn served as a source for Canadian charts of this region.[25]

The success of Kurt caused the German naval weather service to organize a second Labrador station.[26] This time, however, the MWD planned to site the station further south, closer to the 54th parallel. Weather developments at such a more southerly site than Kurt, they reasoned, would better reflect the climate in the maritime zone surrounding Newfoundland. The new station was designated WFL 30, code named "Herbert." By June 1944 the task was assigned to the large, type IXC/40 submarine *U-867*, which had just been built. In Stettin it took aboard the latest automatic station, manufactured by the firm of Siemens-Schückert Werke AG. After final preparations, *U-867* departed Kiel in September 1944 for what turned out to be an ill-fated maiden voyage. Scarcely eight days out of harbour, the brand new vessel radioed a series of emergency signals: both diesels out of service, batteries dead, heavy enemy air cover, storms, and massive seas. No one survived. By the end of the month the bodies of the skipper and six crew members were found on the Norwegian shore.

In August 1944 Grand Admiral Dönitz directed two of his submarines, *U-541* and *U-802*, to operate in what submariners called "the tube" (*der Schlauch*) in the mouth of the St Lawrence. Because other U-boats had been there before them, Dönitz regarded the area as an ideal choke point for picking off convoys moving seaward from Montreal and Quebec City. Both skippers recognized this "gap" as an ideal position for controlling all shipping moving up and down the river. The convoy gathering point was Île du Bic. Though apprised of the tactical situation, the skippers of *U-541* and *U-802* were utterly surprised by the underwater dimensions of their challenge. As Kurt Petersen, skipper of *U-541*, recalled:

> We didn't know the water conditions in the Saint Lawrence, but very quickly ascertained that conditions obtained here which greatly favoured our sojourn ... Water density increased markedly from the surface to greater depths. On the top was the fresh water from the river, and below the salty Atlantic water had probably been pushed underneath it by the tides ... And in order to lie dead with stopped engines at any particular depth of water, we didn't need to employ the so-called hover gear which would keep us at a pre-set depth by automatically flooding and pumping a few litres of water. The gear was used in operational areas in order to conserve energy and to make as little noise as possible when we were, so to speak, at "lurking stations." Here, however, the water layers bore us; they also had the great advantage that antisubmarine vessels could detect us either not at all, or only with great difficulty, for the sonar sound waves were deflected at varying depths. Thus we felt as secure as in the bosom of Abraham.[27]

From 1944 on, most submarines were equipped with the snorkel. This "snort," as it was called in Allied circles, was a tube projecting upward from a submerged submarine and reaching just above the surface of the water. Fitted with non-return valves, it enabled submariners to suck air into the U-boat while simultaneously expelling the diesel exhaust. No longer having to surface in order to run on diesels and charge batteries, the "submersible torpedo boats" of the early war years now operated virtually as "complete submersibles." Thus both *U-802* and *U-541* could do what their predecessors of 1942 could not have done. *U-541*'s "tour" of the Gulf of St Lawrence and River provides a striking example. By mid-September 1943 the boat had snorkeled to within nine miles of Île du Grand Caouis (Grand Cawee), a barren granite island off the northern shore between the mouth of Rivière Pentecôte and Rivière aux Rochers. Fixing his position by a sight of Cawee Light and Egg Island (Île aux Œufs) some fourteen miles to the southwest, the skipper began his inshore patrol. Remaining submerged, he encountered no navigational difficulties. Indeed, he comfortably practised coastal pilotage. Thus by following the twenty-fathom line and relying on occasional periscope checks of local features, he skirted all dangers and operated in some cases to within one and a half miles of shore in complete safety. The crew's experience in this Canadian war zone prompted the submarine's medical officer to write the rollicking doggerel poem "Der Schnorchel," which the crew sang to the tune of the internationally famous hit song of the day "Lili Marlene":

On the St Lawrence
In the narrow spout
Along comes an old snorkel
And lifts its cheeky snout.
Right where the big fat merchantmen steam
That's where we wait to pierce the screen
With you, Lili Marlene.[28]

Nor did the squib stop short of caricaturizing Canadian defence and extolling the virtues of a "peacetime" war zone with easy navigation:

Now the Convoy Commodore's
Growling in his beard:
The local strategic thinkers

Are really rather weird!
But when we're sent on ops once more
We'll want nothing better than a St Lawrence shore
With you, Lili Marlene!

Invariably, oceanographic conditions gave the tactical advantage in Canadian waters to the German submariner. As Canada's director of Operations Division advised the chief of Naval Staff in November 1944: "The schnorkel has reduced the possibility of sighting to a minimum, and the only way in which the positioning of a U-Boat is likely to be fixed is by offensive action on his part."[29] That proved to be the case. Whether sparked by *U-806*'s sinking of HMCS *Clayoquot* on Christmas Eve 1944 in the Halifax approaches or by *U-1232*'s savaging of Boston-Halifax convoy BX 141 off Chebucto Head in January 1945, or indeed, whether triggered by *U-190*'s sinking of HMCS *Esquimalt* near the Sambro Light Vessel the following April, Canadians could never respond until the Germans had first initiated action. And even then, oceanographic conditions frustrated the concerted efforts of Canadian naval forces to destroy the submarines. *U-806* was counterattacked and pursued by naval task force W-12, consisting of twenty-one ships; *U-1232* was hunted without success by the experienced Escort Group EG-27; Escort Group EG-28 hunted in vain for *U-190*. As *U-190*'s skipper later explained, "Canadian search groups worked excellently [but] could not assume that instead of diving in deep water I went into twenty-four metres and lay low."[30] In any event, sonar technology of the day did not permit detection. All the U-boats escaped submerged by dead-reckoning navigation, though in one case pursued almost to the point of exhaustion – by more naval vessels than exist in the whole Canadian Navy today.

Navalists in Canadian defence circles have pointed out that German incursions into Canadian waters argued for the need for a large, highly trained anti-submarine fleet. That, however, would be a clear case of standing prepared to fight the previous war. If such were the measure, Canada has failed to draw appropriate conclusions. An alternative view emphasizes the importance of understanding and documenting the seas in which one operates, and it underscores research and development. The real lesson learned, however, has been taken to heart.

Certainly, when the experience of German operations in Canadian waters was presented to the Federal German fleet's annual "Historical-Tactical Conference" in the Naval School at Mürwik in 1987, this was precisely the view expounded by naval participants. German mariners navigating Canadian waters had penetrated an enticing frontier that beckoned them to undertake bold explorations. Each mission was a voyage of discovery. Here they encountered unusual surface and subsurface characteristics, as well as a wide variety of geographic and hydrographic phenomena. They had found terrain and navigational features that challenged their seamanship and impressed them with the extraordinary range and variety of Canadian waters: Arctic conditions, ice navigation, the swift descent and dissipation of fog, the suddenness of storms, the effects of fresh water outflows meeting ocean tides, unexpected currents and density layers, and of course, the distinctive features of shallow inshore waters. In short, German mariners had grappled with the many facets of a maritime region inviting closer scientific examination. Triggered by its role in guarding the northern flank of NATO through the use of surface and submarine forces in the shallow waters of the Baltic, the Federal German Navy grasped the analogy. From a tactical and acoustic perspective, the oceanographic conditions of the Baltic were found to be as "horrendous" as off the Atlantic coast of Canada and in the Gulf of St Lawrence.[31] The right lesson for Canada, Germans argued, was Canada's actual post-war response: accepting the challenge to broad-based research and technological innovation, especially bathythermography and hydrographic services. That was the wisest conclusion to draw from the German incursions.

# 9

# *Canadian Technical Advances in Hydrography since 1945*

DAVID H. GRAY

The year 1970 saw the completion of the nine hundred and eighteen topographic maps at a scale of 1:250,000 that cover all 10 million square kilometres of the emerged land mass of Canada.[1] In addition, 92 per cent of the thirteen thousand 1:50,000 map sheets have now been published. By contrast, there are under one thousand nautical charts, at all scales, of Canada's 7.7 million square kilometres of land covered by fresh and salt water. As of 1982, only 45 per cent of the country's waterways were surveyed adequately for current requirements, and only 15 per cent of Canadian Arctic waters had been surveyed.[2] Many parts of the submerged coastal areas have only been surveyed at the reconnaissance level. Twenty years later the statistics have not changed significantly. On the emerged land, air photos and satellite imagery point to man-made changes and erosion, but at sea the bottom is invisible, and water erosion and sedimentation occur far faster than on land. These changes require a constant program of resurveying. To do this – meeting the requirements of ever-larger ships, charting new routes, and surveying the submerged land masses even once – means that the Canadian Hydrographic Service (CHS) has to be constantly innovative, in order to get the "most bang for the buck."

This chapter reviews the ships and boats used to gather the soundings, the processes of measuring the depth of water, the tides and currents, some innovative platforms, the tools needed

CSS *Acadia*. (Maritime Museum of the Atlantic, Halifax)

to position the ships, the survey points on shore (and in space) that are an adjunct to those positioning systems, the presentation of the results in the field, and finally, publication of the end products. Each of these phases has gone through upheavals in the past half-century; so the story starts with the status quo that previously existed in each case and outlines the technological advances.

In 1945 the only ships available to CHS were *Acadia*, built in 1913, and *Wm. J. Stewart*, built in 1932. In the immediate postwar era, without money for newly built ships, the fleet was augmented with converted minesweepers (*Kapuskasing*, *Fort Francis*, and *Marabell*) and coastal defence vessels (*Ekholi* and *Parry*). In 1956 *Baffin*, Canada's largest hydrographic ship to date, was completed. She had capacity for six launches, two helicopters, and two barges. She also housed geophysical equipment such as gravity meters, magnetometers, and deep ocean winches. She was followed in 1963 by *Hudson*, in 1967 by *Parizeau*, named for the regional hydrographer on the west coast, and in 1968 by *Dawson*, named for the founder of the Tidal and Current Survey. The gigs under oars that had been used for sounding work were replaced in the new ships with motorized launches with capabilities that improved over the half-century.

The process of taking soundings, both in ships and from their boats, developed steadily. A hand-heaved lead was the traditional method of sounding. In the 1930s a ship, using a lead-line and a donkey engine to haul the line, could sound waters of less than 40 metres while travelling at five knots. The depth could also be measured by pressure sensor (valid up to 180 metres) or by a harpoon sounder.[3] Many seconds passed between throws of the lead, whereas the acoustic sounder, developed from submarine detection equipment, could take several soundings per second in shallow waters. Acoustic sounding in Canada started in 1929 with simply a hammer hitting the bottom plates of *Acadia* and the use of a stopwatch-type device by the recorder. The introduction of a wet-paper recorder was the first improvement, the second being the replacement of the hammer with a magneto-striction device. At first, acoustic soundings were greeted with skepticism, but they are well accepted today. For years, each and every lead-line sounding was shown on a field sheet (i.e., plan of survey results), but with continuous profiles, there was first the manual and then the automated selection of soundings. The shallowest were important for navigation, but they put a bias into the presentation, so the deeps were needed, as well as extra soundings that showed a change of gradient.[4]

For years, analogue recording of the soundings on miles of rolled paper was the only way of preserving the data. During the 1960s CHS groups worked hard to find ways to digitize the data as they were observed and record them in various mediums: paper tape, then nine-track magnetic tape, cassette tapes, and finally, bubble memory. In 1969 CHS organized a survey near Tadoussac, Quebec, to test many development ideas in order to improve work capacity. High-speed launches (including hovercraft), innovative equipment such as paravanes, digitizers – the precursors of software – and procedures such as simultaneous parallel sounding with station-keeping measurements between launches using audio ranging equipment developed by National Research Council were all tried.[5]

During the 1960s CHS tested side-scan sonar deployed from a towed submersible "kite" to understand the hydrography between the sounding lines, but it provided only an assessment of the bottom roughness and identified hazards that might have been missed between lines of sounding.[6] Knowing that a hazard existed meant that time was not wasted looking in areas where there were no hazards.

To provide actual data for the area between tracks of the sounding launch, an experimental solution was an array of sounders aligned on an aluminium TV antenna mast laid horizontally between two Laser sailboat hulls and the whole rig pushed by a launch or small ship.[7] This technique provided 100 per cent coverage of the bottom under the swept area of the arrays. A more robust solution proved to be booms deployed from the catamaran *F.C.G. Smith*. As many as forty-two thousand data points are logged per minute, of which only 0.01 per cent are shown on large-scale field sheets, and even fewer on a chart.[8]

In 1988 CHS faced a serious problem – too much data! The Department of Surveying Engineering at the University of New Brunswick (UNB) established an Ocean Mapping Group to tackle this vast data problem and has continued as a centre of academic excellence in hydrography. D.E. (Dave) Wells, a one-time member of Navigation Group at CHS Atlantic and then a professor at UNB, supervised the development of prototype software, which evolved into CARIS HIPS (Computer Aided Resource Information System – Hydrographic Information Processing System) to sift out the most significant data.[9]

An alternative solution is the present-day multibeam sounder that transmits a fan of narrow sonic beams athwartships. The return echoes at various slant angles from the vessel's track are detected on an array of sensors. The knowledge of the water layers of different acoustic velocities (functions of temperature, salinity, and pressure) can be computer modelled, and the curvilinear path of any oblique acoustic energy can be converted to a depth.[10] CHS assessed the depth accuracy and the effect of the ship's heading and attitude (roll, pitch, and yaw), all of which needed accurate determination, on the position error budget.[11] The *Frederick G. Creed*, which is a Small Water-plane Area, Twin Hull (SWATH) ship of twenty metres, is a very stable, fast ship, capable of working in open water because of its novel design; it uses this equipment.

Soundings from ships' boats underwent similar development. By the late 1930s, in place of swinging a lead from an oared gig with a crew of six, CHS employed an acoustic sounder in a gas-engined open boat, two hydrographers to measure the angles simultaneously, a coxswain to drive the boat, a gas engineer, and a single seaman to operate the sounder. By the 1950s boats were enclosed, with a cabin for most of the length, leaving the coxswain outside and the hydrographers standing up through a hatchway with only head and shoulders exposed. Later, however, once the launch was positioned electronically, one hydrographer monitored the equipment and advised the coxswain. A left-right indicator provided real-time information that the boat was left or right of a particular range-positioning line from a station or a particular hyperbolic pattern line.

The 1950s launch was a wooden displacement hull-form about eight to ten metres long and capable of eight knots. The long keel and skeg protected the propeller. In the 1960s it was replaced by an eight-metre-long, twenty-knot (with no load) planing hulled, fibreglass boat. The propeller and rudder were exposed, which was disastrous in shallow water. CHS even tried jet-powered boats, but they proved difficult to manœuvre at slow speeds.

Traditionally, an area is sounded in parallel lines running perpendicular to the anticipated contours of the bottom. Earlier, with sextant sounding, the direction of travel was totally at the hydrographer's discretion. Later, with range-bearing and two-range positioning, the easiest sounding lines to follow were circular arcs, and with hyperbolic systems, one of the two sets of hyperbolae. To steer other courses was more difficult and was avoided if possible. Today, with computer recording of data, hydrographers can preplan their survey to have sounding lines exactly the right distance apart and going in the desired direction.

Shoals are the headache of any survey, but very important since they define the limiting draft and limiting width of a navigable channel. Shoal areas are identified from the soundings already on the field sheet, notes taken during the soundings, or breakers located from the triangulation points. Hydrographers drop a buoy from the launch and then run a star pattern of lines passing the buoy from different directions to locate the minimum depth.

Sounding information must be integrated with tides and currents. Since charts show the least depth to be expected, CHS has to know what was the height of tide at the time of each depth measurement. For the hydrographer, there are two associated problems. The first is the development of accurate and reliable devices for recording the height of tide. This becomes the data from which tidal information must be extracted. Winds, atmospheric pressure, river discharges, oceanic currents, seiches, and earthquakes all affect the observed tidal or inland water heights. Mathematically, tides can be thought of as the sum of various cosine waves of differing amplitudes, periods, and phase offsets, each expressing the effects of the earth's spin, earth's orbit around the sun, and moon's orbit around the earth.[12] Water tends to "travel" along the coasts and up or down the rivers and bays, giving different heights and times of high and low water at adjacent places. Harmonic constituents that govern the predictable factors must be extracted from the observed data. They in turn can be used to generate tide-table predictions. Thus the second problem was the development of the mathematical analysis and prediction methods that would be used for the tide tables.

As late as 1955, to record water heights, CHS was still putting a seaman ashore to spend the day reading the tide staff on a twenty-minute basis in support of field operations. At permanent tidal stations, the typical self-recording tidal station of the 1900s had been a float in a stilling well linked to a clock-driven drum to record the tide level. Telegraphic exchanges or the meridian transit of the sun regulated the timing of the clock.[13] As part of the International Geophysical Year (1957), Gerhard C. (Gerry) Dohler, chief of Tides, Currents and Water Levels before becoming director of Marine Cartography, initiated a year-round tide-gauging program in the Arctic.[14] Several other types of gauges have been used: a pressure diaphragm gauge that operated even with ice-cover and a bubbler system that measured the pressure needed to force an air bubble out the end of a submerged plastic tube.[15] Tidal observations have shown that some areas of the Arctic have tides as high as those occurring in the Bay of Fundy, while other areas have very small amplitudes. Understanding these effects is important to navigation and for predicting pollution transport and climatic conditions.

By the mid-1950s, graphical techniques determined the harmonic constituents by the analysis of short-term tidal records. In 1959 strip chart recorders (unattended for several days) began replacing the older drum units, which had required daily maintenance. In the early 1960s CHS developed a scaling device to digitize the strip charts with output to punch paper tape, and by 1962 semi-automatic computer processes determined the harmonic constituents. CHS was one of the first maritime agencies to employ digital computers to prepare and publish the data in its tide tables. The development of the Ottboro tide gauge in 1967 was a major step forward in the collection and subsequent processing of tidal records because it produced results directly on punch paper tape, thus eliminating the digitizing. This unit could be installed offshore because it was the marriage of the Ott strip chart recorder and a pressure bellows sensor.

During the 1960s the Tides and Water Levels development group installed a few experimental telemetry gauges to provide real-time water-level information. The system grew to a full network of forty telemetry gauges in the Great Lakes and St Lawrence River.[16] This enabled the forecasting of water levels of the Quebec City–to–Montreal section of the St Lawrence River based on records of the recent past, long-term cyclical behaviour, and meteorological conditions throughout the drainage basin. Ship captains could now load their ships at foreign ports for the predicted depth of water at their Canadian destination. In the 1980s CHS Pacific Region developed a low-power, portable tide gauge suitable for use at temporary sites,[17] and at the same time the Quebec Region developed the TeleMar digital tide gauge for permanent sites.[18] An excellent computer program was written to analyze tidal parameters, and the supporting documentation made the program internationally popular.

The Tides and Water Levels section was involved in other projects. Dohler was instrumental in the adoption and implementation of the International Great Lakes Datum (1955), which was essential for the St Lawrence Seaway construction.[19] He also played a prominent role for more than a decade in the international tsunami warning system around the Pacific Rim – a non-navigation role for CHS but one that provided protection for life and property.[20]

The technology for measuring currents has also changed. Following the demolition of Ripple Rock in Seymour Narrows in 1958[21] and to assist navigation, CHS tidal officers started a program of measuring currents in narrow passes. The movement of drift poles (wood lath with counterweight) was measured by observation. At night, flashlights in sealed milk bottles floated through range lines set up on the shore. As a first development, W.S. (Stan) Huggett used a high mast and long boom to position a current meter in exactly the same spot yet allow traffic to use the channel being measured.[22] This device won a Public Service suggestion award. On the east coast the current survey in the Northumberland Strait first used self-recording current meters in 1958. These Hydrowerkstaetten paddlewheel meters were bottom-moored to obtain continuous records for up to twenty-four days. On the west coast Huggett developed this technology over several years as he extensively surveyed the tide and current in the Strait of Georgia. His results led to the successful *Current Atlas of Juan de Fuca Strait and the Strait of Georgia*.[23] Experiments around Vancouver Island between 1973 and 1980 showed complex currents, which varied in strength and direction as a function of depth and also showed the effects of storms.[24] (These factors also presented very real challenges to Huggett in the development of his meters.)[25] Now, emergency-measure agencies can anticipate the effect of oil spills.

The construction of Cold War warning systems, a global awareness of the continental shelf, and a need to exert Canadian sovereignty variously brought pressure to examine previously uncharted Arctic waters. In the north the hydrographer faces unique challenges of severe weather, short field seasons, and remoteness. Innovative platforms had to be developed for the work.

The Polar Continental Shelf Project (PCSP) was established in 1958 to provide logistical support across all government departments working in the Arctic, and within a year hydrographers had blasted holes in the ice at one kilometre spacing to lower a lead-line from a Lucas Sounding Machine. The next year they drilled holes through two metres of ice.[26] Before long, an echo sounder with the transducer coupled to the ice by a film of oil quickly replaced these techniques. R.M. (Mike) Eaton headed the Hydrographic Section of PCSP from 1959 to 1963, and during the latter year he used a helicopter-towed transducer in Hell Gate and Cardigan Strait, two channels off Jones Sound that provide access to the inner part of Canada's high Arctic islands.[27]

In 1969 the hydrographic section attached to PCSP identified the need to carry out a hydrographic survey seaward of the Mackenzie River delta, an area of interest for petroleum reserves and oil tanker transportation. To survey this shallow area, it tried a hovercraft skimming over the water and mud flats at sixty knots. The sounder was on a boom that hung over the side of the hovercraft, but at high speed the boom caused cavitation that stopped any soundings being transmitted or received.

In 1974, a time of increased interest in domestic petroleum reserves, proposed pipeline crossings between various Arctic islands and from them to the Canadian mainland were surveyed in winter using an echo sounder on a tracked vehicle equipped with a mechanical auger to remove the snow. The transducer was contained in a rubber bag full of antifreeze solution for good acoustic bonding. Metal spikes hydraulically rammed into the ice replaced the process of oiling the ice surface and eliminated snow removal.[28] The technology was so successful that it was installed on helicopters.

The 1990s saw the testing of the electromagnetic Towed Inflight Bathymetry System (TIBS™), which became operational in 1992. It is deployed by helicopter flying at one hundred kilometres per hour and can measure depths to fifty metres through sea ice and turbid seawater by measuring the electromagnetic reflection coefficient of the seawater over a range of frequencies.[29] Thus there was the possibility of longer field seasons, as well as a continuous profile, instead of spot soundings.

Canada has known for twenty years that it will need to delineate the outer limit of the juridical continental shelf in accordance with the United Nations Convention on the Law of the Sea. To cover the offshore subarctic areas in a dense pattern from the continental shelf break to seaward of the foot of the continental slope, CHS first conceived the idea of running unmanned launches arrayed abreast of the mother ship.[30] A better platform was DOLPHIN (Deep Ocean Logging Platform with Hydrographic Instrumentation and Navigation), built by International Submarine Engineering Ltd of Port Moody, British Columbia, which was an unmanned semi-submersible with a snorkel, radio mast, sounding equipment, and diesel engine, all controlled from the mother ship. Being deep in the water, it was unaffected by waves and proved to be a good platform.[31] But the tests, although successful, were abandoned because of decreasing budgets.

Different experiments were also conducted over a period of years to mount a laser light source and sensor in an aircraft. The laser light pulse reflects off the water surface and also off the

Tracked vehicle for sounding through Arctic ice. The positioning system antenna is attached to the white box on the mast, and the hydraulic ram for the sounder's transducer is mounted on the front of the vehicle. (Canadian Hydrographic Service)

bottom, usually as a weaker return. The difference in time is a direct function of the depth of the water. A graph of signal strength against depth provides an indication of the limiting depth of the LIDAR (Light Detection and Ranging) in those waters. So, if there is only the return signal from the surface, then the water is probably greater than the limiting depth. This equipment is arranged to sweep an arc under the aircraft, providing many "hits" per second; thus little water is unsurveyed. The problem with this technique is that shoals and areas of no returns have to be examined by launch. Fortunately, the advent of the Global Positioning System (GPS) removed many of the inadequacies of conventional positioning systems, making LIDAR attractive for surveying ice-free areas in the Arctic and in other inshore areas where the water is clear.[32] Terra Surveys Ltd has used the technology, called LARSEN in honour of the captain of the *St. Roch*, to survey the shallow, southern Northwest Passage, where non-ice-strengthened cruise ships are starting to transit.

While sounding techniques and launch propulsion were developing, the electronic equipment on ships for determining position evolved. During the 1920s, radio direction finding (RDF) became available throughout most of the world. Captain Frederick Anderson, chief hydrographer from 1925 to 1936, and *Acadia* tested the use of the radio directions against sextant fixes along the Eastern Shore of Nova Scotia in 1919 and found that RDF was not accurate enough for hydrographic use. Yet it remains the *in extremis* aid to navigation on ships. CHS became involved in this technological development on two fronts: the verification that signals from stations travelled in straight (actually geodesic) lines and the portrayal of the stations not only on Mercator-projection charts but also on special, wartime, gnomonic projection charts, where the direction lines – geodesics – are straight lines for the plotting of enemy radio broadcasts from the mid-Atlantic.

The gyro compass was the first electronic piece of equipment installed on ships for the betterment of hydrography. To expedite the recharting of the Gulf of St Lawrence and possible future work in Hudson Bay, where the magnetic compass is sluggish, a new gyro compass was installed on *Acadia* in the spring of 1928. Similar units were added to *Lillooet* (1930), *Wm. J. Stewart* (1932), and *Cartier* (1934). Before the Second World War, sounding beyond the sight of land involved considerable dead reckoning where course steered was important. The first gyro compasses were quite problematic and required constant tinkering.

Although the electronic engineers were at work between the wars, most of their efforts do not show up for the hydrographers until after 1945. The Decca navigation system was developed by the British during the Second World War and had its inauguration on the eve of D-Day. The system worked as continuous waves of different frequencies being broadcast from four transmitters: Master (6–*f*), Red (8–*f*), Green (9–*f*), and Purple (5–*f*). The signals from Master and another station produced a set of hyperbolae of constant lane values. Thus mariners could position themselves on specially prepared charts that contained these lattices. Because the signals were continuous, reflections off the ionosphere, known as "skywave," limited the range to about 240 nautical miles and could cause the "odometer" lane counter to slip one or more lanes.[33]

For Decca, a single velocity was assumed for each "colour" pattern. The Royal Canadian Navy assisted CHS in determining "fixed errors" between what was actually observed and what was calculated for locations around the coast. These fixed errors allowed mariners to improve their positioning accuracy. Later, Paul Brunavs, a diligent geodesist from Latvia, compared results with formulae developed by J.R. Johler, who espoused the theory that radio waves travelled at non-uniform velocities.[34] In 1969 Brunavs and Wells mounted a study of Decca frequency transmissions in the Gulf of St Lawrence and proved that Johler's formulae were correct and accurate.[35] A similar test was carried out in Amundsen Gulf in 1973 at the request of petroleum exploration companies which again substantiated their use over water, but proved that radio waves over ice were subject to slower velocities and could not be predicted from just the salinity of the ice or its thickness.[36]

Commencing in 1955, the Decca Master was installed on a hydrographic ship and the two other transmitters on selected shore points, in order to improve the position fix geometry. The disadvantage of two-range Decca was that those parts of the radio spectrum were dedicated to that one ship. With the distances to the two shore points, the position at sea could be determined by trilateration.[37]

Loran-C worked in the same part of the radio spectrum as Decca but as pulses of high energy. The signals were timed so that there was a constant time difference at a stationary point between the signals received from different stations. The high power and the fact that the receiver tracked the beginning of the pulses meant that the signal travelling along the earth's surface arrived prior to the skywave, allowing a twelve-hundred-kilometre range. Again, the publicly available system displayed the results as hyperbolic lines of constant time differences. Mike Eaton, for years head of the Navigation Group at CHS Atlantic, and A.R. (Tony) Mortimer, an ex-mariner turned hydrographer at CHS Pacific, mounted several surveys, which proved that signals travelled at one non-uniform velocity over land and another over seawater, thus confirming Johler's formulae.[38] Because the transmitters were hundreds of miles inland, if one assumed only a seawater velocity, then there would be systematic errors in position up to several miles when the repeatability (capability to return to the same readings) was one-quarter mile or better.[39]

Another part of the radio spectrum used by hydrographers is the medium frequencies, such as used by Loran-A and Hi-Fix. Some people regard Eaton's early work in Hi-Fix as more important than what he did with Loran-C.[40] With Hi-Fix he studied the velocity, locking constants, and site selection and wrote a manual that became the international standard. He verified Hi-Fix velocities against Johler's formulae along the Nova Scotia coast in CNAV *Sackville* using accurate positions derived from shore control.[41] These findings allowed CHS to survey accurately up to two hundred kilometres from shore.

A number of manufacturers produced two-range positioning systems in the microwave and radar frequency bands. Only after several years of operation did hydrographers realize that the cause of weak signals was the destructive interference of the reflected radio wave off the water. Consequently, several hydrographers attempted to find solutions; one proposal was the use of a weak signal alarm.[42] The Syledis system, operating in the UHF band, was much less affected by signal cancellation. It also used measurements to more than two points, which was another improvement.[43] CHS incorporated distance-measuring equipment with angle measurements to give range and bearing location of a roving target – a survey launch. Manufacturers successfully incorporated the two technologies into one piece of equipment.[44]

The United States Navy Navigation Satellite System, commonly known as Doppler satellite positioning, provided the next advance in positioning. If a train has a very predictable schedule and you hear the Doppler effect of its whistle, you can determine the moment the train is closest to you and the distance that it is away.

Applying that principle to signals from a satellite passing overhead, you can determine your position. S.T. (Steve) Grant and Dave Wells realized that Loran-C, smoothed by ship's speed log and gyrocompass, gave the trajectory of the ship during the twenty-minute satellite pass; thus they could determine the ship's position at some point during that time to an accuracy of plus-or-minus 150 metres. That Doppler satellite position enabled them to compute the instant of transmission of the Loran-C signals, hence positions between the satellite fixes, to the same accuracy.[45] This discovery developed into the sophisticated BIONAV integrated navigation system.[46] The solution fulfilled another need; namely, the geophysicists who were on board during many of these offshore cruises needed an accurate velocity to correct their gravity measurements. All this gear allowed CHS and the Geological Survey of Canada, Natural Resources Canada (NRCan), to gather bathymetric, magnetic, gravimetric, and seismic data throughout the Atlantic seaboard in comprehensive multidisciplinary surveys needed for the definition of Canada's juridical continental shelf and preliminary data for resource exploration.

The soundings, reduced for tides, are thus connected horizontally to shore, but they still need to be connected to a geographic reference frame, such as latitude and longitude, so that mariners can relate the charted soundings to where they are at that moment. For years, charts were completed from a single survey, with a local astronomic position, one measured length, and an azimuth. Unfortunately, there were great uncertainties in astronomic positions; hence it was preferable to relate surveys to a national geodetic grid.[47] But the geodetic network reached the coasts slowly, so that many areas continued to have local astronomic positions well into the 1950s. The many points of reference for taking horizontal sextant angles were established from these points of origin by interconnecting the points through a series of triangles. However, it was not the general practice to leave a lasting mark as to the physical location of these triangulation points. To add a new survey to previously charted information, there has to be common points of reference. R.J. Fraser, dominion hydrographer from 1947 to 1952, was frustrated when searching for one point north of Lake Superior, which probably prompted him to institute the practice of marking survey points with proper brass discs.[48]

Typical hydrographic shore station. The hydrographer is measuring angles using a theodolite underneath a large tripod and mast, which is used as a sextant sounding target. CGS *Maxwell* is offshore. (Canadian Hydrographic Service)

Before the Second World War these survey points were positioned by triangulation using a repeating theodolite to measure the angles. Each angle was "turned" over and over again, so that it was successively added to what was previously observed. To read both vernier scales, the hydrographer had to move around the instrument and around the legs of the large tripod set up over the survey point. The potential for knocking the theodolite was great. To spend all day at one station was not unheard of. The saviour to this slow process was the Wild T-2 theodolite, introduced into the CHS in the 1960s. The surveyor, looking into a microscope eyepiece immediately beside the eyepiece for the telescope, measured the angle down to one second of arc by means of a micrometer. The measuring method also changed. The observer chose one station as the initial point and read the circle readings at the other points, and inverted the telescope and took readings at the points in reverse order. Then the observer changed the initial circle reading at the initial point and repeated the process three more times, thus reducing the amount of walking around the tripod by over 50 per cent.

South Africans researched and designed a microwave distance-measuring device, Tellurometer, which came to CHS in 1957. The need to form overlapping triangles so that distances could be calculated by trigonometry from a single baseline measured by invar tape was replaced by measuring all the lengths and angles of polygons, a process called "traversing." The time needed to establish the position of the survey control was thus greatly reduced. Helicopter support also reduced travel time significantly. Brunavs, in Ottawa, was not satisfied with the manufacturer's claim for the instrument and so spent weeks testing the equipment on lines of known length in the Ottawa area. When he had finished, he had learned the probable sources for errors (atmospheric, reflection off intervening ground, and the location of the electrical centre) and had devised methods to reduce their effects.

An interesting little device, a hand-held, Swiss-made CURTA calculator, removed the tedium of always doing calculations with seven-place logarithms. With Tellurometer lengths to be corrected for wet and dry bulb temperatures and for atmospheric pressure and then reduced to sea level and for slope, and geographic and UTM positions being computed for each station, these little marvels reduced computing time immensely. Today, with desktop computers that will do all this and more in a faction of the time, it is hard to remember the CURTA's impact on time and accuracy.

During the 1970s the Doppler satellite system provided one determination of an absolute position every few hours. If many satellites were observed over a period of two days or more at a stationary point, then a position accurate to plus-or-minus one metre relative to the others observed during the same period was obtained. It was used for a network of primary points established by the Geodetic Survey Division of NRCan to position its traverse networks previously run between Shoran trilateration[49] points. Since CHS surveys networks were tied to geodetic traverses, it meant that surveyed positions had to be recomputed.

In the 1990s, GPS gave an absolute geographic position in real time, continuously, twenty-four hours per day. With a constellation of twenty-four satellites revolving around the earth, the user receives signals from at least four of them. Each satellite transmits a time-coded message giving its orbital parameters. But the signals arrive after different travel times, and the computer within the receiver determines the position of the user (latitude, longitude, and height, or X, Y, and Z) relative to the centre of gravity, accurate to plus-or-minus ten metres, and a timing correction for its internal clock. Gone is the need for weeks of establishing sextant targets along the coast or erecting fifty-metre masts for Decca transmissions. It is a simple matter to set up a GPS receiver at a known local survey point as a reference station that provides real-time differential corrections to all the roving GPS receivers. By working in a differential mode from a local reference station, accuracy of plus-or-minus one metre is obtainable. The accuracy obtainable by the mariner is far better than most surveys used in the preparation of charts in front of him. Because of this constantly available accurate positioning system, CHS has had to convert the geographic grid on its charts to match that obtainable by GPS. This has required archival research into the compilation files for the charts, the history of coordinate values of survey points, and the comparison of charted positions.[50]

The end result of all this survey work is the presentation of results on field sheets. These are the large-scale plans, often a metre and a half square, that show all the survey results. One hundred years ago they were drawn on heavy paper. But paper, particularly in conditions that may vary from cool and damp to hot and dry, can change dimensions. It was imperative to lay down the grid lines and the most important triangulation stations as fast as possible before the paper changed dimensions. The next improvement, still in use in the late 1950s, was to glue the paper onto a thin sheet of aluminium. This stabilized the dimensions, but the end product was as mobile as a sheet of plywood. Up to the 1960s, field sheets were constructed using the polyconic projection. It was simple to calculate the location of the intersections of meridians and parallels using published tables. However, there was a change to the Universal Transverse Mercator projection in the 1960s, because it is easier to compute

CSS *Baffin* navigates through ice while one of the starboard side launches is off sounding. (Canadian Hydrographic Service)

the positions of features in UTM coordinates than in geographic latitudes and longitudes, an important factor in pre-computer days. Field officers could go to the field with Mylar (plastic) sheets already plotted with the standard UTM grid, and all they had to do was to label the grid intersections and then start plotting their data in terms of the UTM positions.

Early experiences with computers in the field began in 1963 when an automatic plotting system was installed on *Baffin* to plot depth information directly onto a field sheet. Eventually, the equipment was taken off the ship, and parts of it ran for years in the offices at Bedford Institute of Oceanography. G.R. (Ross) Douglas, dominion hydrogapher from 1987 to 1995, oversaw the first digitizing of soundings in 1969 on board survey launches and the processing of the data by computer back at base. Problems included size, weight, and power requirements. Hydrographers were never sure that there would be any data to process after a day of sounding. By 1974 the computer went aboard launches to filter depths and positions, make sure the data was recorded properly, and provide straight line navigation and line-keeping information. Commencing in 1981, bubble memory solved the years of troubles with cassette tapes for a memory storage medium.[51]

Now, with totally automated recording of the field data and the computer plotting of only a selection of the soundings, the preliminary versions of a field sheet are printed on cheap paper until the final version is ready and then plotted on Mylar. In addition, because soundings are collected in a dense grid, including overlapping data sets, computer algorithms are necessary to recognize erroneous acoustic returns, make sounding selections, remove superfluous data, and contour.[52] In fact, things are so sophisticated that the computer file is the record of the survey that is used in chart construction.

Field sheets also incorporate the shoreline, which up to 1930 was all done from the boats. This process consisted of rowing from point to point with a patent log astern and offsetting by estimation distances less than one hundred metres.[53] In 1930 CHS officially arranged with the Department of the Interior for air photographic coverage of parts of the North Shore of the Gulf of St Lawrence.[54] The coastline was traced from the air photos after matching the plotted survey points to its pinprick location on the photo. By the 1970s it was common to request from Topographic Survey a photogrammetric plot of the coastline taken at low tide with infrared images, because they produced the clearest image of that important chart feature.[55]

The culmination of all these soundings, positionings, coastlining, and geographic referencing is the publication of an accurate chart. Charting, too, has seen an evolution. Copperplate engraving was the *de rigueur* method of printing charts prior to the lithographic printing that started making inroads during the 1930s. In the earliest days of CHS the hydrographer was responsible for preparing his "fair sheet," which was his vision of what the chart ought to look like. The copperplate engraver converted his vision into the actual chart, being very faithful to the original fair sheet.

At one time, CHS had over two hundred charts printed by copperplate. The mirror image of chart information was etched into the surface of a smooth copper plate. When ready for printing, the plate was smeared with a thick paste of ink, and the surface wiped clean, leaving ink in the etched areas. Then very costly high-rag-content paper was pressed against the plate. Press runs were typically small – maybe less than fifty. Amendments to the chart for the next edition were made by hammering out the area to be amended, and the plate re-etched. Early Canadian surveys had been sent to England to be printed via copperplate engraving by the Admiralty. From 1909 to 1911, copperplates were engraved in Canada under contract. The Printing Bureau of Public Works took over the copperplate engraving in 1911 and continued until 1921, when the unit was transferred to CHS. About 1914, Canada took over the printing of Admiralty charts of Canadian waters by photolithography.[56] CHS's last copperplate engraver lasted to 1947, before being reassigned.

Lithographic printing started at CHS in 1904 with a chart of Lake Winnipeg by preparing negatives of each colour for the chart. The waxy coating on the glass negative was scraped away with a chisel-like pen, and the glass plate was photographed to make the lithographic printing plate. Corrections were made by filling in the wax surface, and the process repeated. But the danger of breaking the glass plates made the process unpopular.

By the 1950s the hydrographer no longer prepared a "fair sheet" since charts were often an amalgamation of several field sheets. The preparation of the charts became the responsibility of permanent office staff. Surveys done in detail had to be simplified, contours added, and only a few soundings selected. With charts prepared in metric units, starting circa 1970, the bottom was shown by contours and even fewer soundings.[57] Non-metric field sheets needed to be recontoured at the metric intervals – an incentive to do it with computer programs.

A pantograph reduced field sheets to the scale of the chart, and the reduction was traced onto the manuscript.[58] Later, field sheets were reduced photographically. Since field sheets were drawn on polyconic or UTM projections and charts are typically on Mercator projection, field sheets that covered wide expanses had to be cut into segments to fit the grid of the desired chart. The compilers were the ones to make the decision on the sounding selections – the vital portrayal of the bottom.

Once the compiler had prepared the information that was to be presented on the chart in proper location on a manuscript, the draftsperson's job was to draw a black line paper version of the chart, which was photographed. The title block and text were added using lead type. Starting with chart 4368 (St Ann's

Harbour) in 1953, draftspersons prepared the negatives on plastic (a world first) for each of the four colours. They traced the shoreline and contours from the compilation sheet onto scribe coat (a plastic sheet with a wax-like coating). The reproduction staff could then make a peel coat, where the areas between various lines could be peeled off. A peel coat was prepared for the buff of the land, for various percentages of the blue, and for the magenta.

The complexity of making lithographic maps reached its apex at CHS with the printing of the fifth edition of the General Bathymetric Chart of the Oceans (GEBCO). Prince Albert I of Monaco, an enthusiastic amateur oceanographer, headed a committee that had successfully produced the first and second editions. The International Hydrographic Bureau took over in 1928 but was unable to complete the third and fourth editions. G.N. (Gerry) Ewing, dominion hydrographer from 1972 to 1978, agreed to draft the fifth edition and to provide copies to the IHO. This set of sixteen Mercator sheets, two polar stereographic sheets, and one world map was completed in 1984.

Computers have become involved in all aspects of hydrography; probably the most revolutionary is in chart construction, but it was a long time in gestation. Dr A.R. (Ray) Boyle, of the University of Saskatchewan, under contract to CHS, designed a basic system of automatic cartography for CHS in the late 1960s. The automatic drafting system included a Gerber 32 table and a Barr and Stroud lighthead, both driven by a PDP-8 minicomputer, and a semi-manual digitizing unit invented by Dr Boyle.[59] Digitizing was slow because of the care required to follow lines accurately and the computer's speed in accepting new data. From this beginning, CHS started a tedious and laborious process of digitizing field sheets to build up a digital hydrographic database, alleviated by the fact that new surveys were submitted in digital form. It took about six weeks of work to digitize and verify a full-size field sheet.[60] Later, the digitizing focused upon converting manually compiled chart material and then drafting the chart automatically and later still producing electronic navigational charts (ENCs).

CHS established computer-assisted cartography during the 1970s by developing an in-house, Graphical On-line Manipulation and Display System (GOMADS) to handle digital chart files.

GOMADS evolved in a number of ways and became the basis of a number of other systems. In 1981 CHS helped Dr Sam Masry, of the University of New Brunswick, establish Universal Systems Ltd in Fredericton, and the software development, support, and marketing of GOMADS were gradually handed over to Universal Systems under the name CARIS. CARIS is now the main computer software used in many hydrographic offices throughout the world, a good testament to Canadian development.

In 1983 Mike Eaton painted a futuristic picture of a ship being navigated with the assistance of a chart display on a computer screen.[61] It was not just an image of the paper chart; it provided real-time positioning, navigation information, radar target trajectories, tidal predictions, port facilities, docking assistance, and so on. It all seemed unreal, yet twenty years later most aspects are available. Electronic chart files come in two varieties: raster and vector. Raster is just pixel images of the paper chart, whereas the user can query the vector version to find out background information such as symbology. The latter will also maintain the size of symbols as the user changes scale. T.V. (Tim) Evangelatos, who pioneered computer-assisted cartography, was instrumental in many of the developments of the ENC.

Eaton started the CHS Electronic Chart test-bed project in 1984, using the CARIS GIS as the core, and ran it until shortly before it was demonstrated (and well received) during the Norwegian North Sea ECDIS (electronic chart display and information system) tests in 1988.[62] The conclusions from this project formed the basis of IHO technical standard S-52 (Content and Display of ECDIS) and subsequently the International Maritime Organisation (IMO) ECDIS Performance Standards, encouraged by Evangelatos and others.

With raster or vector files of the chart, it was a simple matter to print the chart on a full-width roller electrostatic colour printer. Don Vachon led the investigation into this method and evaluated some of the equipment on the market.[63] The advantage here was that only a few copies need be printed at a time, and chart files could be updated for each notice to mariners. Technical problems addressed were the quality of the paper and the quality of the inks.[64]

The twentieth century ended with an explosion of information becoming available on the Internet. The government of

Canada is no exception. CHS has made available to users its chart indices, list of dealers, tide tables, and notices to mariners (in conjunction with Canadian Coast Guard) and the limits of various levels of Canadian jurisdiction (in conjunction with the Department Natural Resources).

One hundred years ago (1904) the only change in hydrographic surveying and charting over the previous century had been the vast increase in the number of charts of previously uncharted waters; however, large parts of Canada remained uncharted. Seventy-five years ago (1929) saw the first rumblings of technology advancements: acoustic sounders and gyrocompasses. Fifty years ago (1954) marked the true start of a change to every component of hydrographic surveying and charting: sounding, amount of bottom coverage, transport, positioning, field data presentation, chart compilation, chart drafting, printing, and storage. It has been truly a remarkable half-century, which has resulted in the surveying and charting of much of Canada – a process that is not, and never will be, complete.

# 10

# *At Sea with Hydro: William Metcalf and* USS Edisto*'s Arctic Cruise, Spring 1951*

GARY E. WEIR

Thursday – 1 March 1951 – At Sea, Gulf of Maine
1st Day out of Boston
Back on the good old USS *Edisto* again; I suspect I have made more cruises on this ship than just about anyone going. We left the Charlestown Navy Yard at 0930 – right on schedule – and went out to President Roads where we tied up to an ammunition barge to take on ammunition. It was snowing fairly hard the whole time. We got away from the barge shortly after noon and are now steaming NE with a moderate swell coming from ahead. The ship has hardly rolled at all and is pitching only slightly. We'll pass Cape Sable in the morning and start our BT [bathythermograph] operations at that time.[1]

Nearly every year for the first decade of his career, William Gerrish Metcalf left his home in Falmouth, Massachusetts, drove to Boston, and there boarded a ship to the arctic. To document long days spent at sea, to record private professional observations, and to cope with missing his wife, he made a habit of going beyond personal letters and traditional research notebooks. Most of his reflections on these voyages went into detailed diaries or logs, retained for years in his private files.

Born in 1918 and a veteran of the Second World War, Metcalf completed a master's degree in biology at Amherst College in

1942. Four years later, shortly after his discharge from the Navy, he joined the staff of the Woods Hole Oceanographic Institution (WHOI) as a research assistant. He remained with the institution for his entire career, retiring as an associate scientist in 1979 and working thereafter as both a teacher and consultant. An historian's dream, this experienced polar scientist converted his handwritten notebook logs into word-processor files shortly after his retirement and donated the result to the WHOI archive.[2] On more than one occasion in his early career, Metcalf questioned the wisdom of spending so much time away from his family in the far north. Upon reflection, it often did not seem terribly profitable. Through personal experience, he discovered the relatively low position occupied by Arctic survey oceanography in the US Navy's pantheon of priorities. Its Hydrographic Office ("Hydro") emerged from the Second World War with severely limited assets and hydrographic and oceanographic responsibilities beyond its customary work in surveying, navigation aids, and chart-making. As a result, ambitious and rewarding ventures in the Arctic proved the exception rather than the rule.

By 1946, Hydro had a new Division of Oceanography, led by Richard Fleming, a veteran of the Scripps Institution of Oceanography and the wartime University of California Division of War Research. Upon his arrival at Hydro, he quickly learned the frequency with which he would have to go cap in hand to navy superiors and friends in the oceanographic community to meet his division's commitments. Fleming had only recently acquired two attack cargo ships, converted for survey work to replace older, retired vessels – veterans of interwar Hydro surveys of strategically important areas in the Caribbean and the Pacific. He found himself constantly scrambling to assemble basic instruments and to acquire or retain trained people. The rest of Hydro fared no better. Vital resources remained terribly scarce, and the survey work suffered.[3]

In the first decade after the Second World War, the top of the world did not draw the scientific or strategic interest that the subsequent course of the cold conflict with the Soviets might suggest. The Arctic Ocean, for example, did not rank in importance with the Gulf Steam, a significant ocean anomaly very close to the North American coast. Nothing illustrated this discrepancy better than Operation Cabot, conducted in 1950. This carefully planned, well-manned, six-ship investigation of the Gulf Stream in co-operation with the Canadians produced data critical to a better understanding of that Atlantic phenomenon. During this period, Metcalf and his fellow Arctic scientists yearned for the application of Cabot-like scientific assets to their portion of the world ocean.[4]

Unfortunately for their work, highly classified national defence ventures of both limited and long duration frequently diverted Hydro's ships and many of its best people. "Black" projects, such as the development of the SOSUS deep-ocean surveillance system, and other classified efforts, such as the creation of the Distant Early Warning system, or DEW Line, in Alaska and northern Canada, certainly provided regionally specific data that satisfied some hydrographic needs. However, broad survey science missions could not compete with the priority given to anti-submarine warfare or the effort to monitor the short trans-polar routes available to Soviet bombers.

Most major Arctic projects in the first decade after the Second World War traced their roots to defence concerns and resources. In 1946 Project Blue Jay created a weather station and air base at Thule, Greenland. At the same time the American Congress made appropriations available under Project Nanook to fund a multi-part, joint effort with Canada to establish a series of five other weather stations across the northern Canadian archipelago. Metcalf would discover later that these efforts went far beyond communication and meteorology. They constituted initial steps designed to explore the feasibility of the DEW Line. The massive sealift operations from the United States required to support these projects made some of Metcalf's Arctic voyages possible. For the most part, these joint operations included American sealift ships, under the control of the Defense Department's Military Sea Transportation Service (MSTS), as well as US Navy, US Coast Guard, and Royal Canadian Navy icebreakers.[5]

After his spring 1951 cruise, Hydro asked Metcalf to leave home again almost immediately in support of one of these sealift efforts, a logistics and supply fleet headed for Thule. Metcalf agreed but only with the provision that the navy fly him home when the convoy has safely passed through the ice. His log introduction for the episode describes it well:

USS *Edisto* – Arctic Cruise
Baffin Bay – Thule, Greenland
1 June to 10 July 1951
It was probably while I was still on the *Edisto* on the Spring Cruise in 1951 with Tommy Austin [from Hydro] that Tom recruited me for the summer cruise. I was very reluctant to spend another summer in the Arctic after being away so long from wife and family, but Tommy was very persuasive. He told me that a very large convoy of ships was being assembled to spend the summer among the Canadian Islands doing preliminary work in setting up a series of bases for what was to become known as a Defense Early Warning Line (DEW Line)across the Arctic from Alaska to Greenland. This whole matter was fairly hush-hush at the time ... this great convoy of ships could start out in early to mid-June and get right up to Thule, Greenland, from which base they would fan out across the Canadian Arctic.

The *Edisto* was assigned the task of going up to Thule ahead of everyone else to check out the ice conditions, and then along with the Eastwind lead the Task Force of LST's, LSM's, APA's, AKA's, Survey Ships, Salvage Tugs and what have you to the operating area. The Task Force Commanding Officer wanted to have someone along who knew, or was thought to know, enough about ice and ice navigation to help in such matters. The Hydrographic Office was appealed to, and, I suppose to sweeten the pot a little, Hydro was told that there would be ample opportunity for gathering hydrographic station observations along the way.[6]

Given the nature of the expected threat from the north, the air force also invested in a permanent research center, built on a huge free-floating section of fractured shelf ice. Frequently called Fletcher's Ice Island after air force Colonel Joseph Fletcher, whose radar exploration of the ice pack revealed the extent of its thickness, ice station T-3 was created by the Alaskan Air Command in 1952, with A.P. Crary of the US Air Force Geophysics Directorate as its first chief scientist. Moving slowly with the pack ice, the T-3 station conducted or sponsored regular scientific projects supported by civilian scientific authorities as well as the navy and the air force. Between 1951 and 1955, T-3 moved from the north polar regions to a position just off Ellesmere Island.[7]

Without the favour and priority that came from the navy's interest in the defence aspects of oceanography, Hydro's fundamental survey activities survived on a meagre budget and a traditionally close relationship with private institutions. The latter, such as the Scripps Institution of Oceanography and WHOI, recalled all too well the profitable partnership with Hydro before the war that had helped them survive the lean times before Pearl Harbor and 7 December 1941. Hydro worked in partnership with Scripps, WHOI, and others to continue developing aids to navigation and their entire menu of basic services derived from soundings, BT readings, and oceanographic stations made at sea. This continuing co-operative relationship made Metcalf's presence on so many Hydro Arctic voyages possible.[8]

Work in the Arctic suffered from yet another limitation. Even after the advent of nuclear power and the early successful transits made by USS *Nautilus* (SSN-571), USS *Skate* (SSN-528), and other submarines under the North Pole between 1954 and 1960, naval strategic planners did not view the Arctic Ocean as an operational battle space. American submariners did not see themselves waging war under the ice. Even the father of the Arctic submarine, Dr Waldo Lyon, constantly courted the submarine admirals and those in the navy bureaus who controlled the money for Arctic research and the development of polar operational capability. His Arctic Submarine Laboratory in San Diego survived or foundered on his ability to persuade those in authority that knowledge of the far north and capability under the ice would grant the navy an advantage in the Cold War with the Soviet Union. Many times during his long career and in spite of the suitability of the nuclear submarine for under ice operations, Lyon's research and his lab came close to dying. In the end, his work survived only because of the immense respect that American submariners had for his devotion to their operational welfare. In later years the Soviet submarine force helped his cause by regularly positioning their Delta IV and Typhoon strategic ballistic missile submarines under the ice for isolation and protection. American fast-attack submarines had to seek them out to monitor their behaviour as best they could.[9]

Thus it quickly became clear to anyone who went to sea after the Second World War, bound for the Arctic with Hydro, that its polar survey hydrography and basic oceanography did not have priority. Nearly thirty years after his spring 1951 voyage on the USS *Edisto*, this poverty of resources, data, and priority remained vivid in Metcalf's mind. Sitting at the word processor to compose an introduction to his 1951 *Edisto* log, this long-time polar survey scientist still felt the frustration:

Falmouth, Massachusetts
August 1988
By the end of 1951, the Navy Hydrographic Office had been attempting to collect high latitude oceanographic data for some five years, and the sum total of the information collected was pitifully meager. I write this with great humility since I was the person most responsible for gathering what data they had obtained over this time period. Mind you, there were plenty of ready-made reasons why this was the case – lack of adequate equipment and lack of field experience by the observers, but most of all a real lack of commitment on the part of the Navy. Year after year they would send forth one or more icebreakers with a list of missions including oceanographic observations, but always the oceanographic work appeared almost dead last on the list of priorities. We who went along as oceanographic observers were always made well aware of the fact that our oceanographic work was to be carried out on a "not to interfere basis." This meant that we could only do our work if the commanding officer simply could not think of anything else to do, no matter how trivial. And any competent commanding officer could always think of *something* that could stand doing.

The various commanding officers I worked under were invariably friendly and for the most part very cooperative, but they always had too many things on their list of "missions to be carried out if time permits," and since none of them had had any training or experience in oceanography or any real understanding of it, they naturally did not give our needs much consideration. In fact, in all my experience as an oceanographer on icebreakers, it was only whatever expertise I developed as an ice observer that impressed any of the COs as being of any use whatsoever.[10]

With nearly annual Arctic voyages to his credit in the 1950s, Metcalf quickly became one of the most experienced survey scientists operating in the ice. His particular interest centred on the origins and nature of water transport from the polar regions to the larger ocean basins. The Arctic Ocean and the dynamic processes in Antarctica supplied the major divisions of the world ocean with much of its nutrient-rich water via current action and upwelling. The still incomplete knowledge of these conditions kept Metcalf at sea.

In his account of each voyage and in observations based upon memory added later by way of introduction and epilogue, Metcalf's comments paid particular attention to the people involved in polar survey oceanography during the Cold War, the quality of their professional effort, and the human and environmental challenges they faced. He worked for WHOI and with Hydro, watched them in action at sea, and reflected on the realities of professional and shipboard life in a world before sophisticated modelling, satellite networks, and personal computers.

Before concluding his career, William Metcalf emerged as one of the premier field scientists and polar observers in the United States, with extensive experience in Antarctica and in the far north. His credits included Operations Highjump and Nanook 1 and 2, as well as the equivalent of years at sea above the Arctic Circle on icebreakers and research vessels.

In March 1951 he joined his good friend and Hydro Technical Service chief Thomas Austin on board the Burton Island Class icebreaker USS *Edisto* (AG-89). The scientific program for this voyage to Greenland and the Norwegian Sea included sorely needed general bathymetric research around Denmark's gigantic island and in the adjacent waters.[11] Not only did great gaps exist in the basic oceanographic knowledge of that region, but the navy also realized that this area provided one of the only outlets to the Atlantic Ocean for Admiral Sergei Gorshkov's growing Soviet submarine force.

Just a few months earlier, a group of naval officers and scientists led by Rear Admiral Francis Low and tasked with defining the most potent maritime threat to the post-war United States had submitted their report to the chief of Naval Operations, Admiral Forrest Sherman. Their findings suggested that German technology captured by the Soviet Union and the United States during the war could produce submarines more potent than any the United States had ever faced. The obvious egress routes from Russia's northern ports and bases on the Kola Peninsula included the Norwegian Sea and the choke points in the North Atlantic around Greenland and Iceland.[12] The Hydrographic Office now needed to complete long-neglected surveys of this and other strategically important regions. Concerned about

Soviet submarines in particular, the navy sent USS *Edisto* to sea in the spring of 1951.

Metcalf had a keen eye for observation and a remarkable ability to critically evaluate both himself and others. After years of experience, he also understood that oceanography at sea required strong professional purpose and a large measure of flexibility. Many different kinds of people had to work together in a confined space and in potentially hazardous climate conditions on a task usually defined by others. Thus his observations on the first day out from Boston naturally included a initial take on his shipmates.

Thursday – 1 March 1951 – At Sea, Gulf of Maine
1st Day out of Boston
We have the oddest set of civilians aboard this thing you ever saw in your life. Most of them have never been to sea before and are thus complete strangers to this way of life. I have the suspicion that some of them are in for the shock of their lives.

Tommy Austin has brought along quite a collection of people from Hydro including an eight-man team of Navy BT observers – two chiefs and six enlisted men. This team looks excellent, and it will be quite a relief to have a lot of help in the BT program. Always in the past I have been a member of a one or two man team! When the "Boss" – Tommy Austin – comes along, things get done right!!

... We have set up a watch system that will go into effect tomorrow morning.

Tommy is a tremendous workhorse and will saddle himself with three times his proper load if I don't stop him ... However, with the help of the Navy team, things seem to point towards a better cruise than ever before.[13]

The Navy Department had established the Division of Oceanography of the Hydrographic Office on 29 January 1946 with Reserve Wave Lieutenant Mary Sears of WHOI as the first officer in charge. Richard Fleming took over in April 1946 and organized his division into four sections: Marine Geography, Technical Service, Program Development, and Oceanic Development. Thomas Austin joined Hydro as chief of Technical Service soon after returning from Operation Crossroads, where he had helped the navy monitor the oceanographic effects of two nuclear detonations.[14] WHOI director Columbus Iselin hired Austin and sent him out to the Pacific with a rather large team, including physicist-oceanographer Allyn Vine and marine biologist Gordon Reilly. Among their many achievements, Vine would later create the deep submergence vehicle that bore his abridged name, *Alvin*, and Reilly would leave Yale University to assume a leading role in creating a world-class oceanography program at Dalhousie University in Halifax. Austin would also distinguish himself in spite of the debilitating effects of polio, which plagued him after this voyage on the *Edisto*. Metcalf realized that Austin's congenial personality and his personal drive, evident on board the 1951 cruise, would help his friend survive and prosper.

> Tom's paralysis eventually stabilized so that he had good movement of his head and neck and partial use of his hands and arms. By pulling his hands across a desk with his fingers, he was able to grasp a pencil or pen and make notes, and he became quite facile at feeding himself with the limited dexterity remaining. He was, of course, confined to a wheel chair, but he continued to be active professionally and traveled all over the world to scientific meetings. After a number of years in Hawaii (as Director of the new Fish and Wildlife Service Laboratory), Tommy was made director of the Fish and Wildlife Service lab in Miami, Florida, and after several years there, he became Director of the National Oceanographic Data Center in Washington, D.C.[15]

In spite of his confidence in Austin's ability and determination, Metcalf had long ago discovered the constant presence and unfortunate consequences of inexperience on Hydro voyages. The navy applied the best Hydro resources elsewhere. The team on board ship, which often combined naval personnel, naval civilians, and individuals from private research institutions, needed to inform one another and work together. Only by combining skills did they have a chance at the success they all wanted. Early in each voyage Metcalf recorded examples and consequences of the failure to become professionally acquainted, informed, and integrated. Sometimes he did not have to wait too long. He just hoped that the lessons did not come at too high a cost.

Friday – 2 March 1951 – At Sea, North Atlantic Ocean
2nd Day out of Boston
We passed Cape Sable this morning, and as we turned a little more to the north, we began to take the seas more on our starboard beam, so things started to get rough.

The Navy BT team became all eager to go to work, and without checking with anyone they proceeded to lower a BT. I had previously told them that we cannot work on this ship in rough weather, but they are old salts and knew better. Rough weather was not going to stop *them*! So out they went and lowered the BT, and everything got all fouled up at once. They beat up a brand new 450′ BT – one of our special ones calibrated for Arctic work – so it is unusable even though they got it back. They kinked up 400′ of our precious stainless steel BT wire. And they got themselves soaked to the skin into the bargain.

So now they have stopped that until they get word from us, but it was an expensive lesson to learn.

While the damage to the BT and the stainless steel wire proved less an example of inexperience than a failure to communicate as a team, some of the scientific crew did give Metcalf a reason to pause. In writing to his wife, Betty, within the context of his log, Metcalf mentioned the all-important, Hydro-appointed cruise cartographer, H. McDonald Harper.

He keeps up the ship's "plot" and figures the positions of our observations. He's just a kid and sort of naive, but he's a very nice guy. I think this is his first job, and on the first day I was aboard, he asked me which is latitude and which is longitude – and he's aboard as a professional cartographer!! In fact, he was a Quartermaster in the Navy, though I don't think he was in for long or had any sea duty. He is no ball of fire, and he's a hard man to get up in the morning, but he's a nice kid.[16]

Metcalf also regularly reflected on the willingness of men to get out of their bunks or out of their wardroom chairs long enough to do the very hard work required by science at sea. On the *Edisto's* spring 1951 cruise, this otherwise restrained scientist took to employing obscenities when referring in his log to a particularly useless member of the scientific team from

Hydro. He constantly contrasted this person with the hard-working and dynamic Thomas Austin, his friend and admired role model. In most casual and serious comments about those on board, he would repeatedly refer to the willingness of an individual to do the work. The physically demanding nature of science at sea made devotion, camaraderie, and personal discipline necessary virtues. You had to get out of your bunk, face the sleet and ice, lend a hand with damaged equipment, and volunteer to help when a shipmate needed it. If these things did not happen with fair regularity, the voyage would fail to return with any significant results, and the time away from friends, institution, and family would amount to very little. Metcalf could only justify time away from home by achieving publishable scientific results or by accumulating remarkable experiences that enriched his life and refined him as a professional. Thus he had little patience with an individual who would not "pull his weight." In his brief estimates of the scientific team at the end of the spring 1951 voyage, a willingness to work always drew praise from Metcalf and a positive comment about the individual as a sea-going partner. In each log he would provide short descriptions of the people on board, from the ship commander and the chief scientist all the way down the chain of command.

> He is a typical Boston College Irishman – a nice guy with a fine sense of humor. He's a good worker, though not a particularly inspired one nor a driver like Tommy Austin. He has a lovely looking wife, if his pictures of her are at all accurate ... Then there is Jim McGary from Johns Hopkins. He is a big ... likable kid – a chemist. He hasn't been around much, but he is a very pleasant kid and a very good worker.
>
> The other civilian aboard is William G. Metcalf – a tall scrawny character with a moustache who seems to get along with his shipmates pretty well ... and who misses his wife terribly.[17]

As the ship moved north from Boston, the weather briefly intensified, sending the seasick to their bunks and the rest to inside work, reading, or their games of poker and cribbage. For Metcalf, taking some of this downtime to work with the enlisted personnel on the Hydro BT team helped considerably.

When the storm system passed on the fourth day at sea, the collection of bathymetric data began in earnest. Metcalf lamented the loss of the 450′ BT, designed and built by Lloyd Hoadly and Carl Wing at the WHOI workshop. Before departure, Hydro gave permission for the transfer of the new BTs to USS *Edisto*, together with two new Navy winches. Sloat Hodgeson, who calibrated and repaired BTs at WHOI, modified a number of them for use in the Arctic at sub-zero centigrade temperatures. Now they would have to do without one of them.[18]

Metcalf did not have long to wait before the atmosphere once again began to display its temper. A renewed and angry storm quickly destroyed one of the BT booms. An individual standing BT watch would swing the instrument out and away from the side of the ship on the boom before letting it drop with wire attached into the ocean. This prevented the BT from too easily crashing into the side of the vessel and provided for as straight a drop as possible. The new weather system hit *Edisto* in the Labrador Sea as she made for southern Greenland and turned a boom tested for one thousand pounds into a pretzel, tossed equipment across the deck, and ripped whaleboats from their davits. Metcalf described this storm as the worst he had ever endured, in spite of some spine-testing experiences in the Antarctic. He looked forward to hiding from the elements in the fjords at the southern tip of Greenland within the next few days.

Located well up in a fjord near both the southern tip of Greenland and the city of Narsarsuaq, BW-1 provided a home base and laboratory facilities for monitoring the ice flow through the Baffin Bay and Davis Strait. Metcalf and his colleagues came within sight of icebergs and the pack ice as expected, sighting their first iceberg at 59°06′ N, 48°50′ W. Escaping into the fjord on course to BW-1, *Edisto* encountered only open water until ten miles out from the base, and then broken ice to within one mile and solid ice, one foot thick, as they approached the dock. *Edisto* tied up to the pier on 10 March, ten days out of Boston.

The mercurial nature of the weather and the physical environment made Arctic work at once frightening, challenging, and breathtaking. No sooner did *Edisto* leave BW-1 to continue the voyage than Metcalf returned to his BT work. In the course of just the next twelve hours, he tasted a wide variety of conditions as the ship made for its next port in Greenland.

Monday – 12 March 1951 – Grondal, Greenland
12th Day out of Boston
We left BW-1 at 0500 and went down the fjord in fairly thick fog and a flat calm. It cleared at about 1000, and we started taking BTs. It was a calm and lovely morning. At 1100 there was a slight breeze blowing, and when I came on watch at 1200 it was blowing a gale with mountainous seas. Everyone got soaked, but we got our BT observation and then tied everything down ... The wind and seas forced us to slow down, and the 12 hour run took 14 hours. We entered Arsuk Fjord and got to Grondal in a flat calm with a beautiful sunset. What a day.

While voyages such as this one provided ample opportunity for taking BT readings in a soaking rain, they also offered a chance to recall the true reason for pursuing science. The Arctic could both engage the intellect and overwhelm the senses.

Tuesday – 13 March 1951 – Grondal, Greenland
13th Day out of Boston
This has been one of the best days I've ever spent on one of these trips ...

[A]t 0830 Scotty Matthews took Tommy Austin in the helicopter and flew him up the fjord for a ways and let him off on the ice. About 30 minutes later, Scotty dropped me near Tom on the ice about a mile and a half from the glacier front ... The wind was coming down the fjord at about 45 knots off the ice cap, and we could hardly walk against it on the snow patches. On the smooth parts of the ice we were blown about completely at the mercy of the wind.

We cached the food and coffee on the bank of the fjord and set out for the glacier. It's a smallish one, very blue in all shades of that color, and very rugged looking from the distance. Luckily the wind died down a little as we reached the face ...

We finally equipped ourselves with some hand-sized jagged rocks which we could use to cut into the ice to increase our friction slightly. The whole trouble here is that the experience is completely beyond description. I don't think even the most talented writer could do justice to it, and my poor efforts are indeed stumbling – like our progress over the ice.

The unbelievable blue color in the crevasses, the clarity of the ice, the yawning cracks and towering peaks – they are all so far beyond my power to describe them adequately I could weep with frustration ...

Our minds were supersaturated with all that our eyes had seen. Reluctantly we turned back because we just couldn't assimilate any more.

By this time the wind had dropped to zero, and everything was simply beautiful. We worked our way back to where the food was and ate sitting on a ledge under a towering overhanging rock. Melt water, crystal clear, supplied us with drinking water, and we were tired and happy.

Scotty Matthews came after us at 1515 ... I should sleep like a log tonight.

The helicopter served the dual purpose of local transport for scientific and emergency purposes and as an ice reconnaissance platform for the ship's captain. Knowing the ice distribution ahead or the most advantageous fissures in the ice pack made navigation and breaking through the ice much easier. Even with this information, Captain Nicholas of *Edisto* did not welcome the scattered ice or the need to penetrate into the ice pack for the scientific work or the explosives testing that representatives from the navy's Bureau of Ordnance needed to perform on this voyage. Metcalf and some of the other Arctic veterans found the nervous behaviour of *Edisto's* commanding officer rather amusing. They would take turns visiting the bridge to observe his inclination to dash back and forth between the bridge wings in an effort to assure himself that his ship would escape damage.

Handling the vessel in the presence of a considerable amount of ice and making sure that she did not end up frozen fast in the pack required skill and vigilance. With many of the Hydro ship commanders inexperienced in the Far North, Metcalf frequently found himself at sea with very competent officers suddenly confronted by a challenging, unusual, and often awkward situation. Shortly before they made for BW-1, Captain Nicholas penetrated the ice pack to permit the Bureau of Ordnance team to conduct their tests. Placing the personnel on the ice turned into a truly forgettable episode.

Thursday – 8 March 1951 – At Sea, Labrador Sea
8th Day out of Boston
A week ago today we sailed. We were in ice much of today, coming out at 0445, changing course to 310° and getting back in around 0700. Then we went far enough into the pack to satisfy the BuOrd people, stopped for about two hours for them, and then shoved off for Narsarsuaq. We left the ice about 1845.

You should have seen these people when it came to trying to put the BuOrd party out on the ice to measure the noise. They backed and filled, ran their engines one way and then the other, changed sides, fouled everything up and finally got a little rubber inflatable boat with their equipment in it over on the ice by using the ship's crane. Then they tried to swing the men over to the ice with the crane, too, but with the ship bobbing like a cork in this swell, they decided to give up before they killed somebody. So then they tried getting the men onto the ice using the inflatable equipment boat. That meant dragging it back off the ice and over to the side of the ship, and in doing that, they managed to puncture the boat and it deflated. So they mucked around some more, but they finally gave up. This whole fiasco used up some two hours, and I suspect that a competent outfit could have achieved success in something under an hour. I suspect that the Coast Guard, with its greater experience in handling ships and gear alongside like buoys, etc., at sea could have done much better.

Only the experience that many Hydro ship commanders lacked could properly address this problem. It alone would make placing a party on the ice easier, negotiating the pack and the icebergs more comfortable, and career-ending ship damage more avoidable. Metcalf did not find Nicholas's behaviour irrational or unexpected, only predictable and amusing.

One duty that Captain Nicholas had to perform drove Metcalf mad. A party from the ship called on the American consul in Godthaab, Greenland, when the vessel tied up there for two days, beginning on 17 March. Responding to a number of requests and strong suggestions, Metcalf joined the group for lunch at the American consul's residence. Everyone then went on to dinner at the house of the Danish governor. For Metcalf, all of this seemed awkward, formal, and out of place for a research cruise. Recording his thoughts in his log the next day, he easily remembered why he had never considered a career as a naval officer.

Sunday – 18 March 1951 – Godthaab, Greenland
18th Day out of Boston
We will sail tomorrow morning at 0800 – thank goodness! The guests started pouring aboard at 1100, and we had a buffet dinner and a movie in the Ward Room at noon …

As far as I was concerned, the entire day was a waste of everyone's time, and I wish we could have sailed this morning and gotten on with our work. I suppose the Navy has to do this sort of thing wherever they go – show the flag, and all that sort of thing. I guess that's one of the reasons I never had any desire to be in the peacetime Navy. At least during the war, this type of foolishness played little or no part – or it didn't out in New Guinea where I was, anyway!

On 23 March 1951 *Edisto* rounded Cape Farewell, at the southeastern tip of Greenland, en route to the Denmark Strait in very rough weather, only to emerge into scattered ice and much calmer conditions on the island's eastern coast. At that point Metcalf and the Hydro team initiated the hydrographic stations scheduled for the trip. From the winch and over the side went miles of cable with a variety of measuring and sampling devices, including Nansen bottles, attached at predetermined depth intervals. When the equipment was retrieved shortly after, the onboard team took great care to record the temperature readings, to preserve the samples, and to log all the other characteristics of the water column. Very often, two casts went into the ocean, one for shallower water and another going much deeper. A double cast would usually take just over two hours in freezing but calm sea conditions. *Edisto* experienced just that as she proceeded up the east Greenland coast, completing two hydrographic stations during the course of the day. By the time Metcalf retired to his bunk on 24 March, they had five stations completed and the data recorded, including the necessary temperature readings from the protected thermometers on the Nansen bottles – unusually fine progress given the swiftly changing atmospheric conditions characteristic of the area. Metcalf wondered in his log just how long their luck would hold. As he worked with the Hydro personnel to complete these stations, ship personnel continued to take BT measurements around the clock.

They worked furiously while the weather co-operated in a way that made Metcalf nervous. He had never encountered a run of excellent working conditions before so close to the Arctic Circle. He finished hydrographic stations 8 and 9 on 27 March, and the WHOI-Hydro team successfully completed all the data collection as well as some of the water chemistry. The Nansen bottles not only provided fixed temperature readings for predetermined ocean layers, but they also captured water samples that revealed the chemical composition of the water in particular depth layers. Scientists performed some of the chemical analysis on board ship, including tests for salinity and oxygen content. This kind of chemistry, along with the temperature readings, permitted Austin and Metcalf to compute a water density profile for each location. Combining all of the stations would allow them to draw some conclusions about horizontal water movement. Plotting the data to demonstrate these conclusions also formed part of the on-board work, especially on the homeward-bound leg of the trip.

As he completed these stations, Metcalf crossed the Arctic Circle for the fifth time by sea. He would later make eight Arctic crossings by air while in Skull Cliff, Alaska, working on a classified project related to the creation of the DEW Line. One day earlier, *Edisto* entered the Denmark Strait, and Metcalf completed the day's last station scheduled for 27 March in scattered ice roughly sixty miles off the Iceland coast.

The next day, well into the Greenland Sea and south of Jan Mayen Island, the stations continued. The *Edisto* crew discovered that the newly installed navy winches could not take the strain of both work and the elements and needed constant attention. In addition, taking data in this region with instruments attached to and lowered by cable presented a regular threat of loss and damage. Over the next decade, the advent of regular collection of data by submarines, prototype deep-submergence vehicles, sophisticated buoy systems, and floating ice stations would begin to replace some of the cable-based techniques used so regularly since the First World War. While making excellent progress with their scientific program, including a supplementary effort at taking ocean bottom cores for WHOI, Austin and Metcalf occasionally ran into very familiar difficulties.

Wednesday – 28 March 1951 – At Sea, Greenland Sea
28th Night out of Boston
This completes four long weeks away from home. Time is moving fairly rapidly as we are keeping very busy. We finished Stations 10, 11, 12 and 13. If anyone had told me before this trip started that we would get 13 out of 13 stations on schedule, I would have laughed in his face.

The weather is beautiful, but Tommy's luck ran out on his station, #12. The bottom bottle slipped down [from its fixed position on the cable while deployed] and smashed against the hydro weight, and a precious unprotected thermometer was crushed. The protected thermometer seems OK, and Tommy didn't think the bottle was damaged, but when I went out to do Station #13, I tried to use it, and it wasn't working. So I had to break out a new bottle which delayed my station a little. Then, since the station came at chow time, Tommy had figured out the depths for my station while I ate. So I lowered the bottles according to his lay-out, dragged things on the bottom, lost part of the hydro weight and scooped up mud in the bottom bottle. Fortunately the thermometers were not hurt ... I should have mentioned that at my Station #11, we got a good bottom sample with a Phleger corer.

On the thirty-first night, USS *Edisto* picked up Jan Mayen Island on the radar, reached the turning point of the spring 1951 voyage, changed course south and west into the North Atlantic, and headed for Boston, making stations all the way. Having a ranking oceanographer from Hydro such as Thomas Austin on board doubtlessly helped keep the technical team on task and also prompted Captain Nicholas to find more time for hydrographic and oceanographic work than he might have otherwise. Metcalf constantly celebrated the numbers of stations accomplished and the team's ability to keep remarkably close to the scientific program timetable for the voyage. The return leg took them east and south of Iceland into the Irminger Current, a branch of the Gulf Stream moving west and south toward New England.

As they sailed, Austin and Metcalf kept the data collection effort on track. With the advent of the Irminger's warmer water, the weather began to change again, and the winds picked up along with the waves. At some of the stations the sea became rough, and the cable paying out from the winch on the stern took on an angle relative to the ship that indicated just how high the waves rose and how deep the vessel dropped into troughs. In favourable conditions the cable or "wire" assumed an angle of forty degrees or less. During Austin's effort to accomplish station 26, the angle rose quite a bit, and he spent a good deal of time on board but under water.

Tuesday – 3 April 1951 – At Sea, North Atlantic Ocean
34th Night out of Boston
It was bound to happen sooner or later. Our string of consecutive stations came to the end this morning. The barometer went down to 29.10 inches

and was still dropping when we were due to stop for Station #25. The wind was kicking up a big sea, so we passed up the station. We finally made it this afternoon where station #26 should have been.

This afternoon's weather was fine with a big sea rolling, but the temperature was warm, and the sun was shining. It was typical Gulf Stream weather – which makes sense since we were in a branch of the Gulf Stream system, the Irminger Current, which bends back to the west, south of Iceland.

By evening it was again blowing hard, and Tommy made part of a cast for Station #26 with a 60° wire angle in a 40 knot wind. They were taking water over the side, and things were quite bad, so they quit.

I'm due for #27 before breakfast, but we got under way from Tommy's station at reduced speed, so mine will be delayed.

The weather seemed to reflect their high homeward-bound spirits, for the wind continued blowing furiously and the waves towered, conditions only briefly punctuated by short periods of relative calm. Austin took station 30 at 0500 on 5 April in a dead calm. Five hours later Metcalf took number 31 in a raging wind, with rain and sleet blowing in his face and a fifty-degree wire angle. By the time Austin returned to do 32 at 1500, he started with ninety-degree wire angles and soon had sea- and rainwater all around him up to his armpits. He abandoned the effort shortly after he started.

On 6 April USS *Edisto* began its final run toward Boston. Metcalf and Austin made the last two hydrographic stations at 0200 and 0930 along the edge of the pack ice, and the ship turned for home. Metcalf's positive feeling about the voyage went beyond the presence of Thomas Austin and collecting more significant data. In an aspect of oceanography not in the current limelight, he now had to his credit a challenging research cruise that had completed all but two of its hydrographic stations and provided excellent data for post analysis and publication. He had recently been promoted to research associate at WHOI, and the science accomplished on board USS *Edisto* confirmed his ability and helped to establish additional professional credibility.

Monday – 9 April 1951 – At Sea, Grand Banks
40th Night out of Boston
We slowed down enough last night so that now our ETA Boston is 0800 on Thursday, the 12th. I sent a radio dispatch to WHOI requesting

a truck for that time. Our work is finished, our Homeward Bound Report is finished. From here on in we are just going along for the ride!

The success of the voyage rested largely on the presence and determination of Thomas Austin and the possible defence applications of the scientific results. Having a senior Hydro official on board gave the hydrography and oceanography the authority it needed to rise to the top of the dreaded list of tasks the captain could address on a "not to interfere basis." This voyage did not change the dilemma in which Arctic scientists found themselves; it just demonstrated the possibilities and kept Metcalf in polar work for a few more years.

Two shorter voyages, one later in 1951 and another in the early part of 1952, sent him looking for possible alternative fields of study. Returning to the low priority routine after the excellent results of the voyage with Austin seemed to make up his mind. Arctic work did not stand out as a premier field of interest at WHOI, in spite of the accession of Edward "Iceberg" Smith to the directorship to succeed the retiring Columbus Iselin. A retired US Coast Guard admiral, Smith had helped to set up the iceberg patrol in the North Atlantic as part of the response to the loss of RMS *Titanic* in 1912. He also led some of the first serious exploratory efforts at sea above the Arctic Circle in the 1920s and 1930s, including the notable Marion Expedition.[19] Unfortunately, he did not lead WHOI into Arctic studies, a prospect that Metcalf might have welcomed.

Seriously beginning to evaluate the wisdom of continuing with Arctic research, Metcalf suddenly found himself invited to take a look at how the other half lived. His friend Gilbert Oakley spent some time at WHOI as captain of the institution's research vessel *Atlantis* and also served as port captain before joining the staff of the US Air Force Cambridge Research Center (CRC) at MIT in Boston. CRC managed Lincoln Laboratories in Lincoln, Massachusetts, just west of Boston, a facility devoted to Air Force defence research.

Early in the summer of 1952, Oakley mentioned to his superiors at CRC that Metcalf had the Arctic experience they sought relative to a project of interest to the air force. The CRC Summer Study Group, led by Dr Jerrold Zacharias of MIT, planned to take a few weeks during the summer of 1952 to bring together individuals with every sort of Arctic experience to address a question posed by Air Force planners. Approached by Zacharias, Metcalf received permission from Edward Smith of WHOI to participate in the summer study. After a quick background investigation and the granting of a "secret" clearance, the WHOI Arctic specialist took the train from the old station in Woods Hole to Boston to meet with the CRC leadership. Metcalf quickly found out that the summer study would focus on the viability of the proposed DEW Line. Years later he recalled,

> At the time, this work was all very secret, but what it amounted to was that the Defense Department was interested in the possibility of putting into a place a network of bases across the North American Arctic and Greenland to serve as an early warning system for the defense of the USA and Canada from a Soviet attack by air. This quickly was nicknamed the DEW Line, the initials standing for Defense Early Warning. Actually, much of the exploration among the Canadian Islands by US icebreakers over the past several years under the guise of setting up weather stations had been to look into the feasibility of setting up this DEW Line. There were relatively few people who had much recent experience among the Canadian Islands, and Gil Oakley had suggested that maybe my experiences aboard the icebreakers would make me a useful member of the Summer Study Group.[20]

Metcalf now better understood the reasoning behind the many icebreaker voyages he had joined, the rationale for the logistics and supply priorities, and the frequent difficulty in persuading the ship commanders to spend more time with oceanography and hydrography.

For two or three days each week, Metcalf would work as part of the summer study, collecting as much chart and navigation information as he could on various potential radar sites in the northern Canadian islands and writing up commentaries on their accessibility by air and sea at various times of the year and on the physical conditions that supply ships and personnel might encounter in trying to equip and supply these DEW Line bases. He spent a great deal of time on the train to and from Woods Hole and during the workweek as an overnight guest at his father's home in Belmont.

In discussing and debating the problem on every plane, from physical location to the relative merits of radar versus acoustic detection of incoming Soviet aircraft, the summer study group reached the conclusion that a site already existed which would provide the perfect location for a series of tests. Anticipating a greater volume of air traffic using polar or near-polar great circle routes across the Pacific after the war, the air force had already built a 625-foot-tall skeleton tower as a Long Range Aid to Navigation, or "Loran," station – a first step in extending that valuable navigation tool into the far north. The Loran tower stood at a location just a few miles southwest along the Alaskan coast from Point Barrow, the site of Pet-4, or Naval Petroleum Reserve Facility Four. The tower, already in place, made the ideal test structure for various radar and acoustic listening devices proposed as DEW Line detection methods.

Once agreement came concerning the nature of the tests and the role the results would play in determining the nature of the DEW Line, Metcalf discovered the true reason for his invitation to join the study.

> Eventually, it was decided that the way to find out just how much could be heard for how far in the Arctic was to go to the Loran site at Skull Cliff and conduct a study for which the tower would be equipped with a lot of very highly sensitive acoustical equipment, and various types of planes would be flown in the area at various distances, altitudes, speeds, etc., under a variety of weather conditions. MIT would undertake to round up a few experts in acoustical work, and they wondered if I would like to go along to help. As nearly as I could make out, my role was more or less that of being a handy-man who would be called upon to keep the scientific party, totally unfamiliar with Arctic conditions, from doing something stupid like walking off a cliff into the Arctic Ocean or wandering about the tundra and getting lost or freezing to death because their minds were so occupied with more important things like decibels and sound waves.[21]

Already in the process of re-evaluating the wisdom of remaining in Arctic physical oceanography, Metcalf viewed the invitation as an opportunity to use his environmental expertise, to reflect on his scientific future, and to experience a project with true priority. The latter had rewards he had never experienced.

> Actually, it sounded like it might be an interesting experience, and I was becoming sort of fed up with the endless series of icebreaker cruises

where the Oceanographic Program was always of low priority. I was becoming a first class cribbage player as a result of those cruises, but I didn't feel that I was learning much oceanography or contributing much to the field. Furthermore ... MIT paid fabulously high wages compared with what WHOI was putting out in those days, so the temptation was probably as much financial as scientific, I must confess.[22]

Metcalf's work at Skull Cliff earned him an invitation to join CSC permanently as an Arctic specialist at a salary that WHOI could not match. However, by that time he had already made up his mind. He wanted to do physical oceanography, make a contribution to the field, spend more time at home, and link his work more closely with that done by his colleagues at Woods Hole. He would not stop working with Hydro, but he wanted something new. As he put it in the epilogue to his Skull Cliff log,

That same winter, [WHOI's] Val Worthington had spent several months with a Navy project using the "Ice Island, T-3" as an oceanographic platform, having been involved the previous year in a project on which they took hydro stations through the ice from a Navy R4D (C-47 aircraft) at several locations in the Arctic Ocean. So between us, Val and I had the most Arctic experience among the WHOI scientific staff, and we found ourselves frequently being consulted about Arctic matters. I think that if either of us had wanted to do so, we could have developed a fairly strong Arctic Program at WHOI, but both of us felt that there was equally interesting work to be done in other parts of the world which were more comfortable to work in.[23]

William Metcalf continued his Arctic work through the end of the 1950s before shifting his attention to South America, the Caribbean, and the Atlantic equatorial undercurrent. His participation in the 1957–58 programs of the International Geophysical Year kept him in the north for a time, as did some of his work with the Russians on the Polymode Project in the 1970s. However, unfortunate navy priorities, climate, family, and interests at his home institution effectively took him away from Arctic survey oceanography and hydrography roughly three years after the unusually profitable spring 1951 cruise on USS *Edisto*.

# *Notes*

## ABBREVIATIONS USED IN THE NOTES

| | |
|---|---|
| AMN | Archivo del Museo Naval, Madrid |
| BCA | British Columbia Archives |
| BA-MA | Bundes- und Militärarchiv, Freiburg im Breisgau |
| *CAR* | *The Canadian Annual Review of Public Affairs* (J. Castell Hopkins) |
| CLR | British Columbia Crown Lands Registry |
| *DCB* | *Dictionary of Canadian Biography* |
| DHH | Directorate of History and Heritage, Ottawa |
| Marine | Archives nationales (Paris), Archives de la Marine |
| NA | National Archives of Canada |
| OHS | Oregon Historical Society |
| PRO | Public Record Office, London |
| RGAVMF | Rossiiskii gosudarstvennyi arkhiv Voenno-Morskogo Flota |
| UKHO | United Kingdom Hydrographic Office, Taunton |
| WHOI, DLA | Woods Hole Oceanographic Institution, Data Archives and Library |

## INTRODUCTION

1 That Cook carried chronometers on his second and third voyages is well documented. However, to say that he himself

"benefited" from the development of the station pointer may be problematic. Andrew David has suggested that Cook may have used horizontal sextant angles on each of his three major voyages. Certainly, the technique was known. Murdoch Mackenzie's account of his new instrument to plot horizontal sextant angles, the station pointer, was published while Cook was away on the second voyage. This instrument is not included in the list of instruments provided to Cook for the third voyage. However, during his later life, the instrument was certainly used by other surveyors.

2 Glover, "Early Trading Voyages in Hudson Bay," 96.
3 Chappell, *The Tangier Papers of Samuel Pepys*, 126–9.
4 Hydrographic Department, Admiralty, *Ocean Passages for the World*, 61.
5 Canada, Department of Fisheries and Oceans, *Sailing Directions: Labrador and Hudson Bay*, 10, paras 63 and 64.
6 In writing this paragraph, I am indebted to Dr A.E. Collin, dominion hydrographer, 1967–72, and Mr G. Ross Douglas, dominion hydrographer, 1987–95, for their comments.
7 Meehan, "Oceanography," 12.
8 Taylor, "The Seas in the Seventies," 90.
9 Foster and Marino, *The Polar Shelf*, 11, 9.

## CHAPTER ONE

1 Heidenreich, *Explorations and Mapping of Samuel de Champlain, 1603–1632*, presents the best, most complete assessment of Champlain's achievement.
2 See "Lac Superieur et autres lieux ou sont les Missions des Peres de la Compagnie de Iesus comprises sous le nom d'Outaoüacs," from *Relation de ce qui s'est passé de plus remarquable ... en la Nouvelle France ... 1670 et 1671*. It was probably drawn by Fathers Claude Dablon and Claude Allouez.
3 M.W. Burke-Gaffney, "Boutet de Saint-Martin, Martin," DCB, 1: 119.
4 Pritchard, "Early French Hydrographic Surveys"; Bedini, *The Pulse of Time*, 34–41, clarifies the claims of Galileo and Huyghens concerning the application of the pendulum regulator to traditional clockwork.

5 See Soble, *Longitude,* 27–8, 37–8, for an excellent account of the history of the longitude problem.
6 Olmstead, "The Voyage of John Richer to Acadia in 1670." See also Pritchard, "French Charting the East Coast of Canada," 119.
7 Pritchard, "Early French Hydrographic Surveys," 126–7.
8 Deshayes's record of his work during the summer of 1686 survives and has been summarized in Pritchard, "Early French Hydrographic Surveys," 128–32.
9 M.W. Burke-Gaffney, "Franquelin, Jean-Baptiste-Louis," DCB, 2: 228.
10 Pritchard, "French Charting the East Coast of Canada," 120.
11 Delanglez, "Journal de Louis Jolliet allant à la découverte du Labrador, 1694."
12 Marine, série 3JJ, 274, liasse 1, contains routes and distances of the coasts of English colonies taken in 1699 from a Dutch printed chart.
13 Marine, série B2, 140, f. 111. Deshayes's chart was published by Nicolas de Fer in 1702. A copy is reproduced in Fillmore and Sandilands, *The Chartmakers*, 15.
14 Marine, B4, 32, ff. 297–8, Captain of *Le Profond.*
15 Hahn, *The Anatomy of a Scientific Institution*, 59–63.
16 Ibid., 63.
17 Jones, *The Figure of the Earth*, provides a readable account of this controversy and its resolution.
18 Marine, 3JJ, 259, liasse 12, *Capitaine de vaisseau* Forant.
19 Neuville, *État sommaire des archives de la Marine*, 147.3, 392n3.
20 E.g., Tramond, *Manuel d'histoire maritime de la France*, 370–1; and Jouan, *Histoire de la marine française*, 83.
21 Conlon, "La Condamine the Inquisitive," 362.
22 Russo, "L'enseignement des sciences de la navigation," n3.
23 Pritchard, "Early French Hydrographic Surveys," 133.
24 James S. Pritchard, "Testu de La Richardière, Richard," DCB, 3: 620–2.
25 Joseph Cossette, "Bonnécamps, Joseph-Pierre de," DCB, 4: 76–7.
26 J.S. Pritchard, "Chabert de Cogolin, Joseph-Bernard de," DCB, 5: 175–6.
27 See Chabert de Cogolin, *Voyage fait par ordre du roi en 1750 et 1751.*

28 *Atlas maritime*. See also Bellin, "Remarques sur les cartes du 'Neptune François.'"
29 Hahn, "Les observatoires en France au XVIIIe siècle," 655; Woolf, *The Transits of Venus*, 32.
30 Marine, C2 41, interface, 161–2, "Officiers entretenus pour le service de la marine les quels ne sont attachez à aucun port," 1 avril 1761.
31 James S. Pritchard, "Pellegrin, Gabriel," DCB, 4: 619–21.

## CHAPTER TWO

1 Known to the Spanish as the Battle of Cabo Santa María, after the cape lying to the east of Cape St Vincent.
2 This officer sometimes signed his full surname as Alcalá-Galiano. However, as Malaspina always referred to him simply as Galiano, this form of his surname has been used throughout this article.
3 Kendrick, *Alejandro Malaspina*, 28.
4 For the full text of this letter and accompanying plan, see David et al., *The Malaspina Expedition 1789–1794*, 1: 311–15.
5 The Spanish *tonelada* was a measure of both displacement and carrying capacity, for which it has been proved difficult to establish an English equivalent.
6 AMN, MS 583, ff. 8–10.
7 OHS, Malaspina Papers, MS 2814/45.
8 AMN, MS 2,355, ff 3–3v.
9 AMN, MS 427, ff. 18v-19.
10 AMN, MS 583, ff. 42v-43.
11 AMN, MS 427, ff. 6v-7.
12 Gonzáles, "Instrumentos del Real Observatorio destinados a la Expedició Malaspina," 232.
13 AMN, MS 1,826.
14 Cerezo, *La Expedición Malaspina*, tomo 2, 1:274.
15 AMN, MSS 583, ff 15–15v, 427, ff. 12–12v and 634, ff. 65–7.
16 OHS, Malaspina Papers, MS 2814/183, and Bancroft Library, 92/82z, folder 27.
17 AMN, MS 278, f. 44.
18 Cerezo, *La Expedición Malaspina*, tomo 2, 1:325.
19 One Burgos foot = 0·91 English foot.
20 Phipps, *A Voyage towards the North Pole*, 87–98.

21 Ibid., 99–107. Phipps calculated the distance between his two ships by the simple formula: distance = masthead height divided by the sine of the subtended angle, whereas Malaspina adopted a more complicated formula devised by Galiano.
22 AMN, MS 656.
23 Phipps, *A Voyage towards the North Pole*, 153–79, which contains an illustration of the pendulum used by Lyons.
24 AMN, MS 1826.
25 Espinosa y Tello, *Memorias* (1809), 1:190–212.
26 Cerezo, *La Expedición Malaspina*, tomo 2, 1:215. See also Taylor, "Degree of Difficulty," 18–25.
27 Cerezo, *La Expedición Malaspina*, tomo 2, 1:28.
28 For details of this apocryphal strait and the reasons that Malaspina was sent to search for it, see David *et al.*, The *Malaspina Expedition 1789–1794*, 2: appendix 2.
29 John Ellicott (1706?–1772), a London clockmaker, who made a number of astronomical or regulator clocks. For an illustration of an Ellicott astronomical clock, see Howse and Hutchinson, *The Clocks and Watches of Captain James Cook*, 280.
30 George Adams (1750–1795), a London instrument maker.
31 "Behind the bay ... or rather to the South of it, the chain of mountains before mentioned, is interrupted by a plain of a few leagues extent; beyond which the sight was unlimited; so that there is either a level country or water behind it" (Cook and King, *A Voyage to the Pacific Ocean*, 2:347–8).
32 This is the bay where Bodega y Quadra anchored in the *Sonora* during his first voyage to Alaska in 1775 and which he named on 24 August after Antonio María Bucareli de Ursuá, viceroy of New Spain. It is a large body of water in the Alexander Archipelago in 55°13′N, 133°32′W, separating Baker Island and Suemez Island and leading northeast to the west coast of Prince of Wales Island.
33 Malaspina could not have included Meares's survey as he did not receive a copy of the latter's *Voyages* until his return to Acapulco.
34 The southern extremity of Baker Island in the Alexander Archipelago, marking the northern entrance to Bucareli Bay.
35 AMN, Sig. Vistas. Carp. IV (88).
36 "Sketch by Compass of Port Mulgrave," in Dixon, *A Voyage round the World*, facing 170.

37 Espinosa, *Memoria* (1805), intended as a late appendix to *Relación del viage*.
38 Espinosa, *Memorias* (1809).
39 The *toesa* = the French *toise* of 6·3946 feet, which would make Espinosa's height 5,444 metres, slightly below the figure of 5,489 metres accepted today.
40 Espinosa, *Memoria* (1805), 4–6, or his *Memorias* (1809), 1:58–6.
41 Plans 8 and 9 in *Atlas para el viage de las goletas Sutil y Mexicana* (see illustrations on pp. 33 and 35).
42 This shoal, Roca Pamplona, was allegedly discovered by José Cañizares, *primer piloto* in the *Favorita*, who stated that it was "a half-league around, twelve leagues from the coast in 59°02′ and 40°34′ W of San Blas." Arteaga thought that it was "only white wood floating on the sea" (Wagner, *Cartography of the Northwest Coast of America*, 1:194). The Russians also reported a shoal in the vicinity, but no such reef exists. The name, however, has been perpetuated in Pamplona Spur, which has depths of less than 100 fathoms over it and a least depth of 68 fathoms, and juts out some 35 miles from the coast in this vicinity.
43 The nearest that Cook got to Middleton Island was about 50 miles, when he was rounding Kayak Island.
44 Dixon's route is borne out by the chart showing his tracks in 1786 and 1787; see Dixon, *A Voyage round the World*, facing title page.
45 The naming of this island is wrongly attributed to Vancouver in 1794 in Wagner, *Cartography of the Northwest Coast of America*, 2:397. Vancouver did not sight this island in 1794, since the nearest he got to it was about 30 miles when he was off Kayak Island; nor did he mention Middleton Island in his journal. The island is not charted on Vancouver's large-scale chart from Prince William Sound to Mount St Elias (chart 11 in his published atlas), but it is shown and named on his small-scale chart from Kodiak Island to San Diego (chart 14 in his published atlas). The island was probably named after Admiral Charles Middleton, one of the Lords Commissioners of the Admiralty at the time that Vancouver returned to England. It had previously been sighted by Martínez on 19 May 1788 and named Hijosa.
46 As Punta Verde is not shown on any of the surviving Malaspina charts, it cannot be located with any certainty. It may be Cape Yakataga, situated about halfway between Cape Suckling and Icy

Bay, or possibly Claybluff Point, the western entrance point of Icy Bay.

47 Palau, Zabala, and Sáiz, *Viaje científico y político*, 285.

48 An inlet on the south side of Whale Bay on the southwest coast of Baronoff Island.

49 Espinosa and Cevallos did not appreciate the extent of Port Eliza, which was a fourth channel extending to the north.

50 The two measurements would have been made from each end of the base to cancel the effect of any wind.

51 Espinosa, *Memoria* (1805),6–7.

52 Espinosa, *Memorias* (1809), 1:61–3.

53 See 46–7 below.

54 Plan 7 in *Atlas para el viage de las goletas Sutil y Mexicana* (see illustration on p. 44).

55 UKHO, 355/1.

56 Plate 5 in Vancouver's *A Voyage of Discovery*, atlas.

57 UKHO, 228, press 82.

58 Lamb, *George Vancouver*, 1:287.

59 See also ibid., 1:241.

60 Cerezo, *La Expedición Malaspina*, tomo 2, 1:386.

61 For full details, see *Relación del viage*, 44n1.

62 *Relación del viage*, which was published anonymously. There can be little doubt, however, that entries in it relating to astronomical and hydrographic activities derive from Galiano.

63 Kendrick, *The Voyage of Sutil and Mexicana 1792*, 215.

64 Plan 2 in *Atlas para el viage de las goletas Sutil y Mexicana.*

65 Kendrick, *The Voyage of Sutil and Mexicana 1792*, 204.

66 Chart no. 3 in *Atlas para el viage de las goletas Sutil y Mexicana.*

67 Espinosa, *Memorias* (1809), 1:1–224.

68 *Zapiski izdavayemiya Gosudarstvennim Admiralteiskim Departmentom*, 256–70.

69 Sáiz, *Bibliografia sobre Alejando Malaspina*, 223–4.

## CHAPTER THREE

1 I acknowledge the help of archivists, librarians, and curators in the British Columbia Archives, the Maritime Museum of British Columbia, the Canadian Hydrographic Service, the British Columbia Crown Lands Registry, the Vancouver Public Library Special Collections, the Vancouver Maritime Museum, the

Vancouver City Archives, the University of British Columbia Library Special Collections, Map Library, and Department of Geography Library, the North Vancouver Museum, the National Archives of Canada, and the United Kingdom Hydrographic Office. Sandy Sandilands and Ed Dahl have been generous with their knowledge in discussing charting questions.

2 See Gough, *The Royal Navy and the Northwest Coast of North America 1810–1914*, for historical background. For the sequence of surveys, see Sandilands, *The History of Hydrographic Surveying in British Columbia*, 113–31. Glover, "The Challenge of Navigation to Hydrography on the British Columbia Coast, 1850–1930," deals with the application of technical innovations in hydrographic survey. The present chapter investigates the publication of the resulting charts.

3 Lamb, *George Vancouver.*

4 Gibson, *Otter Skins, Boston Ships, and China Goods*; Howay, "A List of Trading Vessels in the Maritime Fur Trade, 1785–1794."

5 The charts are as follows: (1) Part of the N.W. Coast of America [Fitzhugh Sound]; St. Patrick's Bay [Cape Scott, Vancouver Island: both James Hanna, *Sea Otter*, 1786]; Track of the Snow Experiment [Nootka Sound to Queen Charlotte Sound and Queen Charlotte Islands: S. Wedgborough, 1786] (27 March 1789); (2) Scott's Bay [Cape Scott]; Friendly Bay in Nootka Sound [both Wedgborough, *Experiment*, 1786] (4 April 1789); (3) Port Brooks [Kyuquot, Vancouver Island: James Johnstone, 1787] (15 October 1789); (4) Rose's Harbour [Queen Charlotte Islands: Johnstone, 1787] (15 October 1789); (5) Calamity Harbour [Banks Island: Johnstone, 1787] (23 October 1789); (6) Ahouset [Vancouver Island: Charles Duncan, *Princess Royal*, 1788] (17 December 1789); (7) Port Stephens [Pitt Island, Principe Channel: Duncan, *Princess Royal*, 1788] (17 December 1789); (8) Milbank's Sound [Price Island: Duncan, *Princess Royal*, 1788] (24 December 1789); (9) Port Safety on the East Side of Calvert's Island [Duncan, *Princess Royal*, 1788] (29 December 1789); (10) Bay at the Southern Part of Nova Hibernia; Etches-Sound [both Moresby Island, Queen Charlotte Islands: Duncan, *Princess Royal*, 1789] (1 January 1790); (11) Entrance of the Strait of Juan de Fuca [Duncan, *Princess Royal*, 15 August 1788] (14 January 1790); (12) Clioquot or Port Cox

[Clayoquot, Vancouver Island: Robert Funter] (14 January 1791); (13) Wic-a-na-nish's Harbour (Clioquot or Port-Cox) [Clayoquot: Charles William Barkley, *Imperial Eagle*, 1787]; Friendly Cove at the North Entrance of Nootka-Sound; Raft-Cove in Queen Charlotte Sound [both Funter] (27 May 1791). For fuller descriptions, see Wagner, *The Cartography of the Northwest Coast of America to the Year 1800*, nos. 722–3, 725–31, 745–6, and 781–2; and Cook, "Alexander Dalrymple." These plates were not among those acquired by the Hydrographic Office in 1809–10 from Dalrymple's estate (Richards, *A Memoir of the Hydrographical Department of the Admiralty*, 5).

6 UKHO, Archives, Original Document, p24 on 87; illustrated in Hayes, *Historical Atlas of Canada*, 159, map 232.

7 PRO, ADM.1/3522, Admiralty, Hydrographer's correspondence, 1795–1808, 256–329, enclosure in Dalrymple to W.W. Pole, secretary to Admiralty, 10 October 1807.

8 PRO, ADM.1/3522, 322, "Scale 0.5 inches = 1′ Long. River Oregan. 1/2 Sheet. Arrowsmith 1798. £0.1s.0d." A manuscript of Broughton's 1792 survey is in the UKHO Archives: Original Document 229 on Rv. The engraved "Plan of the Oregan River from an actual Survey" (London, 1798) was reissued in 1831 and 1840.

9 Gough, *The Royal Navy and the Northwest Coast of North America 1810–1914*, 14–23.

10 Ruggles, *A Country So Interesting*, 89–91 and catalogue nos. 105c–107c.

11 Ibid., 90 and catalogue nos. 213a, 114c–115c, and 120c–125c. Ruggles omitted many Simpson manuscripts in the UKHO Archives, including two sketches of Fraser River: Original Documents L3939 on Ac2 and L4226 on Ac2. Admiralty charts between the Strait of Juan de Fuca and Dixon Entrance (for example, chart 1922) mentioned in the text and footnotes are described in the table on 64–73.

12 Belcher, *Narrative of a Voyage round the World*, 1:xxii.

13 Ibid., 1:288.

14 Chart 592, withdrawn from use between 1837 and 1839.

15 Cook, "Alexander Dalrymple," 166.

16 Day, *The Admiralty Hydrographic Service 1795–1919*, 334–5.

17 Ibid., 27–9.

18 [Hydrographic Office, Admiralty], *Catalogue of Charts, Plans, Views, and Sailing Directions*, issues from 1825 onwards.
19 Charts 1577 and 1603, issued in 1839.
20 Gough, *The Royal Navy and the Northwest Coast of North America 1810–1914*, 65–6.
21 Charts 1897, 1901, 1906, 1907, and 1910, all issued 1848.
22 Chart 1911, issued 1849.
23 Chart 1917, issued 1849.
24 Chart 1916, issued 1849.
25 Chart 1922, issued 1849 (see note 10 above).
26 Chart 1947, issued 1849.
27 Charts 2067, issued 1851, and 2153, issued 1852, for Beaver Harbour; chart 2512, issued 1856, for Nanaimo.
28 Sandilands, "The History of Hydrographic Surveying in British Columbia," 119; Gough, *The Royal Navy and the Northwest Coast of North America 1810–1914*, 133.
29 Chart 2168, issued 1853.
30 For Inskip's work under Prevost, see Akrigg and Akrigg, *H.M.S.* Virago *in the Pacific 1851–1855*.
31 Chart 2168, issued revised 1856.
32 Hydrographic Office, Admiralty, *Queen Charlotte Islands on Western Coast of North America*.
33 Chart 2426, issued 1856.
34 Chart 2430, issued 1856.
35 Chart 2627, issued 1858. The 1864 reissue revises the survey date from 1856 to 1857.
36 Dawson, *Memoirs of Hydrography*, 134–9; Sandilands, "The History of Hydrographic Surveying in British Columbia," 119–23.
37 Gough, *The Royal Navy and the Northwest Coast of North America 1810–1914*, 142–4.
38 BCA, CM/C2355, is a dry proof of the Esquimalt chart, sent by Richards to Pemberton in 1861 specifically for the addition of land topography.
39 Hydrographic Office, Admiralty, *Vancouver Island Pilot, Part I*. This comprised "Sailing Directions for the Coasts of Vancouver Island and British Columbia, from the Entrance of Juan de Fuca Strait to Burrard Inlet and Nanaimo Harbour."
40 Hydrographic Office, Admiralty, *Instructions to be followed in Writing and Editing Sailing Directions*, issues from 1857 onwards.

41 Charts 585 and 2067, issued 1863.
42 Charts 582, 583, 569, and 584, issued 1864–65.
43 Charts 570, 590, 1916, 716, 717, 589, and 592, issued 1865–66.
44 Hydrographic Office, Admiralty, *The Vancouver Island Pilot … Compiled from the Surveys Made … between the years 1858 and 1864*.
45 Sandilands, "The History of Hydrographic Surveying in British Columbia," 123.
46 Three charts, all at scale 1:73,000: (1) Vancouver Island. Barclay Sound. By Captn G.H. Richards and the Officers of H.M.S. Hecate. 1861. Initialled "G.H.R." Lithographed at the Royal Engineer Camp, New Westminster B.C. (BCA, CM/A910; CLR, 17T1 Charts); (2) Vancouver Island. Barclay Sound. Alberni Canal. by Captn G.H. Richards and the Officers of H.M.S. Hecate 1861 [northeastern continuation to Alberni; no imprint] (BCA, CM/A911 annotated "R.E. Survey Office"; CLR, 15T1 Charts); (3) Vancouver Island. Part of Barclay Sound. By Captn Richards and the Officers of H.M.S. Hecate 1861. [northern continuation to Pipe Stem Inlet and Effingham Inlet] Lithographed at the Royal Engineer Camp, New Westminster B.C. (CLR, 16T1 Charts, annotated "R.E. Survey Office, New Westminster, March 1862"). Bruce Ward and John Spittle, of the Map Society of British Columbia, kindly alerted me to the Alberni Canal and Barkley Sound "charts"; see Spittle, *Maps Printed by the Royal Engineers, New Westminster, British Columbia, 1861–1866*.
47 Charts 611, 602, and 364, issued in 1865. For Metlah-Catlah Bay, see Gough, *Gunboat Frontier*, 178–180.
48 Dawson, *Memoirs of Hydrography*, 168–9.
49 Sandilands, "The History of Hydrographic Surveying in British Columbia," 123.
50 Charts 2449, 2453, 1901, 2426, and 2190, issued 1869 and 1872.
51 Hydrographic Office, Admiralty, *Vancouver Island Pilot, Supplement*. This comprised the "Coast of British Columbia from Queen Charlotte Sound to Portland Canal, including Queen Charlotte Islands."
52 Hydrographic Office, Admiralty, *The British Columbia Pilot*, including the "Coast of British Columbia, from Juan de Fuca Strait to Portland Canal, together with Vancouver and Queen

Charlotte Islands"; [first edition], with *Hydrographic Notice* [1891] and *Supplement* (1895); second edition (1898), with *Supplement* (1899) and *Revised Supplement* (1903); third edition (1905), with *Supplement* (1908) and *Revised Supplement* (1910). Further editions (from 1913) were issued in two parts.

53 Richards, *A Memoir of the Hydrographical Department*, 22–3 and 26–7.

54 Small corrections, though not yet named as such, are known on chart 1922, "Fraser River and Burrard Inlet," from April 1866 (BCA, CM/C2381).

55 The *Hydrographic Notices* issued by the Hydrographic Office were ephemeral and are consequently elusive. They differ from *Notices to Mariners* in publishing the results of surveys or observations of permanent dangers. They were gradually supplanted after 1900 by regular *Supplements* to Admiralty *Pilots*. The most extensive sets of *Hydrographic Notices* from the 1860s are in the Sailing Directions Branch of the UKHO Archives. The practice before the mid-1860s was to amalgamate, re-edit, and republish area collections of *Hydrographic Notices*, and it is not now possible to reconstruct the earlier sequences completely.

56 Hydrographic Office, Admiralty, *Hydrographic Notice, No. 4: North Pacific, Vancouver Island. Rock in Seymour Narrows, Discovery Passage* (issued 4 September 1866, 1 page). This was the fourth notice issued in 1866 for the North Pacific area, a single slip sheet, of which 250 copies were printed.

57 Hydrographic Office, Admiralty, *Hydrographic Notice, No. 10 [of 1867]: Vancouver Island Pilot, Notice 1 (corrected Notice of No. 4, 1866). Rock in Seymour Narrows, Discovery Passage* (issued 29 May 1867, 1 page). This was the tenth notice worldwide in 1867, and the first formally to amend the 1864 edition of the *Vancouver Island Pilot*. Subsequent *Hydrographic Notices* were: *No. 26 [of 1868]. Vancouver Island Pilot, Notice 2. Vancouver Island, North Coast, Suwanee Rock [etc.]* (16 December 1868, 2 pages); *No. 22 [of 1869]. Vancouver Island Pilot, Notice No. 3. Haro Strait, Rock in Roche Harbour* (1 July 1869, 2 pages); *No. 24 [of 1877]: Vancouver Island Pilot, Notice No. 4: North America, West Coast (British Columbia): Queen Charlotte Sound; Sea Otter Group [etc.]* (27 August 1877, 20 pages); *No. 1 [of 1880]: Vancouver Island Pilot, Notice No. 5 [Esquimalt Harbour, etc.]* (14 January 1880,

4 pages); *No. 13 [of 1880]: Vancouver Island Pilot, Notice No. 6 [Esquimalt Harbour, Thetis cottage, etc.]* (8 June 1880, 1 page); *No. 34 [of 1883]: Vancouver Island Pilot, Notice No. 7 [Victoria Harbour, etc.]* (18 September 1883, 4 pages). The *British Columbia Pilot* of 1888 was amended by one *Hydrographic Notice* before the introduction of formal *Supplements* (see note 52 above): *No. 2 of the year 1891: Notice No. 1 relating to British Columbia Pilot, 1888 [Railways, etc.]* (24 April 1891, ff. 15).

58 Day, *The Admiralty Hydrographic Service,* 81.

59 Meehan, "An Outline of Hydrography in Canada,": "Confederation and Hydrography," 13–15.

60 CLR, 26T1 Charts; another example is at the Maritime Museum of British Columbia.

61 Now UKHO Archives, Original Documents series.

62 Meehan, "An Outline of Hydrography in Canada," part II, 14.

63 Chart 3029.

64 Walbran, *British Columbia Coast Names.*

65 Meehan, "An Outline of Hydrography in Canada," part II, 14.

66 Sandilands, "The History of Hydrographic Surveying in British Columbia," 125.

67 Charts 333 and 3127. See Glover, "The Challenge of Navigation to Hydrography on the British Columbia Coast, 1850–1930," 8.

68 Charts 3260, 3333, and 3387 for Johnstone Strait, issued 1902–04. The large-scale charts 3162, 3178, 3271, 3417, 3430, 3443, and 3448, issued 1902–04, belong to the same surveys.

## CHAPTER FOUR

1 The classic study of the Canadian militia is Morton, *Ministers and Generals.* A full-length related biographical study is Haycock, *Sam Hughes.*

2 Michèle Brassard and Jean Hamelin, "Préfontaine, Raymond," *DCB*, 13: 842–3. The *rouges* generally were associated with the Liberal Party, and the *bleus* with the Conservative.

3 Ibid., 842.

4 Ibid., 843, 844.

5 Gauvin, "The Reformer and the Machine," 17; and Brassard and Hamelin, "Préfontaine, Raymond," 844.

6 Linteau, *Histoire de Montréal*, 124.
7 Brassard and Hamelin, "Préfontaine, Raymond," 844.
8 Gauvin, "The Reformer and the Machine," 18; Linteau, *Histoire de Montréal*, 254–5.
9 Brassard and Hamelin, "Préfontaine, Raymond," 844.
10 Wallace, *Macmillan Dictionary of Canadian Biography*, 736.
11 Penlington, *Canada and Imperialism*, 242.
12 Hopkins, CAR, 1902, 232–7 and passim.
13 Schull, *Laurier*, 411–12.
14 Ibid., 413–14.
15 Hopkins, CAR, 1902, 227, quoting Canada, House of Commons, *Debates*, 7 May 1902; Toronto *Globe*, 19 May 1902.
16 Hopkins, CAR, *1902*, 18. It is interesting to note the contrasting assessment by Laurier's biographer in Schull, *Laurier*, 414: "Préfontaine was a lesser Tarte, with all the little man's weaknesses and none of his strengths. There would be less force in the Cabinet now but there would be less trouble. These men [including Sutherland] were manageable."
17 Hopkins, CAR, 1902, 237.
18 Ibid., 145.
19 Melville, "Canada and Sea Power," especially 222–3, but also chapters 6, 7, and 8 passim.
20 "Naval Defence of Canada," Toronto *Saturday Globe*, 20 June 1896.
21 Gimblett, "Reassessing the Dreadnought Crisis of 1909," 36–8.
22 Ollivier, *The Colonial and Imperial Conferences*, 1:153–5.
23 Ibid., 1:161.
24 Great Britain, Colonial Office, *Conference between the Secretary of State for the Colonies*, Colonial Conference of 1902, appendix VII, 261–2. It is interesting to note that this appendix, with its clear statement of Canadian policy, was not reproduced in detail in the official Canadian history of the conferences (see instead the digest in Ollivier, *The Colonial and Imperial Conferences*, 1:208–9).
25 Ollivier, *The Colonial and Imperial Conferences*, 1:165–6.
26 Hopkins, CAR, 1902, 145–6; 1903, 267–8.
27 Montreal *Gazette*, 25 November 1902.
28 NA, RG 42, Marine Branch, B-1, vol. 87, file 19279, "Investigation into the Grounding of the SICILIAN in the Saint Lawrence

River" [hereafter referred to as *Sicilian* file], "Report of the Commissioners," 21 January 1903.

29 Montreal *Gazette*, 26 November 1902.

30 *Sicilian* file: *Times of London*, 27 November 1902; Préfontaine to Sutherland, 20 May 1903.

31 Sandiland, "Hydrographic Surveying in the Great Lakes," 155.

32 Thomson, *Men and Meridians*, 2:209.

33 NA, Laurier Papers, 71652, 71653ff, and 74825–8; Gourdeau (deputy minister) to Préfontaine, 31 March 1903, rough draft of *An Act to Amend "The Public Works Act" and the Act Respecting the Department of Marine and Fisheries*.

34 NA, RG 2, PC 461 (1904), as quoted in Meehan, "An Outline of Hydrography in Canada," chapter 2, 2.

35 NA, RG 2, PC 1200 (1904), as quoted in Fillmore and Sandiland, *The Chartmakers*, 68.

36 Hopkins, *CAR*, 1904, 27.

37 Meehan, "A Chronology," chapter 2, 3.

38 Ibid., chapter 1, 31.

39 Ibid., chapter 2, 8.

40 Thomson, *Men and Meridians*, 2:212–13.

41 Meehan, "A Chronology," chapter 2, 26.

42 Hopkins, *CAR*, 1903, 436.

43 Ibid., 269.

44 Ibid., 329–30, 364ff and passim.

45 Ibid., 118. See also Canada, House of Commons, *Debates*, 10 October 1903, 13659–60, for official announcement of the construction orders.

46 Appleton, *Usque ad Mare*, 80.

47 "Report of the Fisheries Protection Service of Canada," Canada, *Marine & Fisheries Annual Report*, 1902, 268–9.

48 Maginley and Collin, *Ships of Canada's Marine Services*, 88–9.

49 Canada, *Marine & Fisheries Annual Report*, 1904, 97.

50 Canada, House of Commons, *Debates*, 9 October 1903, 13545.

51 NA, MG27, II, C4, Brodeur Papers, Docket No 2, dated for first reading, April 1904.

52 Canada, House of Commons, *Debates*, 21 June 1904, 5267–70.

53 Gimblett, "Reassessing the Dreadnought Crisis of 1909," 41.

54 *Canadian Military Gazette*, 6 June 1905, 4.

55 Toronto *Globe*, 12 July 1905, cited in Hopkins, *CAR*, 1905, 469.

56 "Reports of the Commanders of Cruisers," Canada, *Marine & Fisheries Annual Report*, 1905, 311–12.
57 Chambers, *The Canadian Marine*, 72, 85, and *passim*.
58 NA, Laurier Papers, 103354–6, Gourdeau (deputy minister, Marine and Fisheries) to Préfontaine, 13 November 1905.
59 *Canadian Military Gazette*, 28 November 1905, 7.
60 John Wesley Dafoe had taken over the position in 1901, after serving in various capacities on Montreal newspapers, in which he was sure to have known Préfontaine. See Wallace, *Macmillan Dictionary of Canadian Biography*, 169.
61 Winnipeg *Free Press*, 9 December 1905.
62 Hopkins, CAR, 1905, 468–9.
63 NA, Laurier Papers, 103995, Laurier to Préfontaine (cable), 5 December 1905.
64 Ibid., 104518–19, Préfontaine à Gourdeau, le 20 décembre 1905.
65 Ibid., 104544, Préfontaine to Laurier (cable), 21 December 1905.
66 Hopkins, CAR, 1905, 503.
67 Ibid., 33.
68 Tucker, *The Naval Service of Canada*, 1:162.
69 Hopkins, CAR, 1906, 575–76.
70 Gimblett, "Reassessing the Dreadnought Crisis of 1909," 45.
71 NA, RG 2, PC 21–1453 (1910); Tucker, *The Naval Service of Canada*, 1:141; Meehan, "A Chronology," chapter 2, 25.

## CHAPTER FIVE

1 The enabling act was not passed until 1868; see Thomson, *Men and Meridians*, 2:31; *Statutes of Canada*, 1868, 31 Vic. c. 57. From 1884 to 1891, the ministry was known as the Department of Marine; it then reverted back to the Department of Marine and Fisheries.
2 Thomson, *Men and Meridians*, 2:202. For an overview history of Public Works, see Owram, *Building for Canadians*.
3 Thomson, *Men and Meridians*, 2:37–8. *Statues of Canada*, 1879, 42 Vic. c.7; Owram, *Building for Canadians*, 137, 139.
4 *Statutes of Canada*, 1873, 36 Vic. c.4.
5 Occasionally, the term "hydraulic survey" was used, even though hydraulics refers to the study of the movement of liquids through pipes or other artificial channels. The St Lawrence River survey at the end of the century was sometimes referred to as a

hydraulic survey. See Canada, DPW, *Report* 1899/1900, 23–4; Prelini, *Dredges and Dredging* , 12.

6 "An Act to transfer the management of certain harbours, piers and breakwaters from the Department of Public Works to the Department of Marine and Fisheries," *Statutes of Canada*, 1877, 40 Vic. c.17; *Revised Statutes*, 1885, 49 Vic. c.84.

7 *Statutes of Canada*, 1892, 55/56 Vic. c. 17; Thomson, *Men and Meridians*, 2:205, 212; Appleton, *Usque ad Mare*, 84.

8 Thomson, *Men and Meridians*, 2:217.

9 Wharton, *Hydrographic Surveying*, 60–1.

10 DPW, *Report*, 1909/10, 6.

11 DPW, *General Report, 1867–1882*, xxv–xxvii, 1206–7.

12 Canada, DRC, *Report*, 1879/80, 141; Cowie, *Report ... on Halifax ... Ocean Terminals*, 34–5; a contemporary description of a survey is in Cunnungham, *A Treatise on ... Harbour Engineering*, 43–59.

13 See DRC, *Annual Report*, 1879/80, 141, for a small survey on the St Lawrence River.

14 Prelini, *Dredges and Dredging*, 12, 14–18; Quinn, *Design and Construction of Ports*, 115–17.

15 DPW, *General Report, 1867–1882*, 584–640.

16 LaPierre, "Joseph Israël Tarte," 47–57. Owram, *Building for Canadians*, 160–4.

17 Prelini, *Dredges and Dredging*, 204–16, describes the project at the turn of the century.

18 DPW, *General Report*, 1867, 37; DPW, *General Report, 1867–1882*, 459.

19 DPW, *General Report*, 1867, 37–8.

20 Owram, *Building for Canadians*, 181–3; DPW, *General Report*, 1867, 39–40; *General Report, 1867–1882*, 454–5, 458, 797. The most recent work, completed in 1993, increased the channel depth to 11 metres (36 feet). This 30-centimetre (1-foot) increase enabled vessels to handle an extra 1,000 tonnes (1,100 tons).

21 DPW, *Report, 1867–1882*, 210–90, is a summary of river improvements in Canada in tabular form. See also Canada, DMF, *Canada: Her Natural Resources*, lists navigable waterways; Black, *Extracts Relating to the Navigability of Canadian Inland Waterways*.

22 DPW, *Report, 1867–1882*, 232–3; DMF, *Canada: Her Natural Resources*, 85.

23 Hind, *Narrative of the Red River Exploring Expedition*, 461.

24 Owram, *Building for Canadians*, 150; Barris, *Fire Canoe*; Peel, *Steamboats on the Saskatchewan*.
25 Canada, *Sessional Papers*, 1885, no. 138, 9, 14–17.
26 DPW, *General Report, 1867–1882*, 253. Legget, *Canals of Canada*, 103.
27 Bennett, *Yukon Transportation;* Letourneau, "Grand Rapids Tramway."
28 DPW, *Report*, 1881/82, 115; 1883/84, 130–1; 1884/85, 138–9; 1899/1900, 215.
29 DRC, *Report*, 1895/96, 103, 114, 120; 1897/98, 35, 130; 1899/1900, 35, 203.
30 DRC, *Report*, 1880, 166; 1883, 109, 114; 1884/85, 139; 1885/86, 137; 1899/1900, 200; Angus, *A Respectable Ditch*, 140–2.
31 DPW, *Report*, 1899/1900, 23–4; 1900/01, 243–51; DRC, *Report*, 1899/1900, 231–6, 249–50. Thomson, *Men and Meridians*, 2:203–4; Owram, *Building for Canadians*, 67, 177–9.
32 Thomson, *Men and Meridians*, 2:204; Owram, *Building for Canadians*, 67, 177–9; DPW, "Georgian Bay Ship Canal Report."
33 DMF, *Georgian Bay and North Channel Pilot*; Canada, Ministry of Fisheries, *Sailing Directions*.
34 DPW, *General Report, 1867–1882*, 86–7, 452, 642.
35 Ibid., 210–90, provides a tabular list of harbours-of-refuge.
36 DPW, *Report*, 1869/70, appendix no. 11, 25–62; Canada, *Sessional Papers*, 1870, no. 49.
37 Appleton, *Usque ad Mare*, 82; Thomson, *Men and Meridians*, 2:205–7.
38 Appleton, *Usque ad Mare*, 53–9.
39 Keefer, "Ice Floods and Winter Navigation"; DPW, *General Report, 1867–1882*, 336–44.
40 DPW, *General Report, 1867–1882*, 756–61, 764–89; DPW, *Report*, 1899/1900, 24.
41 *Statutes of Canada*, 1877, 40 Vic. c. 17, sec. 8. *Revised Statutes*, 1885, 49 Vic. c. 84, sec. 8, and 1906, c. 112, sec. 3. DPW, *General Report, 1867–1882*, 1144–283, provides a tabular list of work done to harbours.
42 Owram, *Building for Canadians*, 144.
43 Norrie and Owram, *A History of the Canadian Economy*, 218–21, 234–40, 321–3.
44 DRC, *Report*, 1880/81,19.
45 Cowie, *Report on ... Halifax ... Ocean Terminals*, 34–6. See also *Historical Atlas of Canada*, 3: plates 19, 25.

46 DPW, *General Report, 1867–1882*, 231; Owram, *Building for Canadians*, 217; McGahan, *The Port of Saint John.*

47 DPW, *General Report, 1867–1882*, appendix no. 6; Quinn, *Design and Construction of Ports,* 166–8.

48 *Vervoort,* "Lakehead Terminal Elevators."

49 DPW, *General Report, 1867–1882*, 516–34; Owram, *Building for Canadians*, 142–3.

50 DPW, *General Report, 1867–1882*, 558–66, 568–82.

51 DPW, *Reports and Documents in Reference to the Location of the Line*, 38, 49, 63.

52 Ibid., 51, 53, 56–7, 66–8; Canada, *Report of the Canadian Pacific Royal Commission*, 1882, 3:84–5.

53 DPW, *Reports and Documents in Reference to the Location of the Line*, 34, 50, 57; Berton, *The National Dream*, 266–70.

54 Talbot, *The Making of a Great Canadian Railway,* 315–17. This was the last Admiralty hydrographic survey on the Pacific coast; see Fillmore and Sandilands, *The Chartmakers*, 62–3.

55 Thomson, *Men and Meridians*, 2; de Tremaudan, *The Hudson Bay Road*, 54–6.

56 Canada, *Sessional Papers*, 1910, No. 20d, 4; 1911, No. 19b.

57 de Tremaudan, *The Hudson Bay Road*, 84–9, 95–101.

58 Thomson, *Men and Meridians*, 2:117.

59 DRC, *Report,* 1884/85, 102–5; 1885/86, 102; 1886/87, 104; 1887/88, 115.

60 DRC, *Report,* 1895/96, 103; 1897/98, 35, 130; 1899/1900, 35.

61 DPW, *Report,* 1899/1900, 23–4; 1900/01, 6; Fillmore and Sandilands, *The Chartmakers,* 128.

62 See, for example, DRC, *Report*, 1895/96, 129.

63 *Historical Atlas of Canada*, 2: plate 47.

64 DPW, *General Report, 1867–1882*, 663.

65 Denis, *Electric Generation*, 123.

66 DPW, *General Report, 1867–1882*, 349, 368–70. One of the most devastating twentieth-century floods occurred in the Lac Saint-Jean area in July 1996, but it was due to an aberrant summer storm, not spring flooding.

67 DPW, *General Report, 1867–1882*, 368.

68 Ibid., 536–56; DPW, *Report*, various years, 1900–04.

69 Thomson, *Men and Meridians,* 2:103–7; DOI, *Annual Report*, 1885, xxiv, 20.

70 *Statutes of Canada*, 1898, cap. 35, "An Act to amend and consolidate the North-West Irrigation Acts of 1894 and 1895";

DOI, *Report*, 1908/09, part VII: "Forestry and Irrigation," 18–19, 87–9; Canada, *Sessional Papers*, 1910, no. 133.

## CHAPTER SIX

1 I am greatly indebted to the late Captain E.H.B. Baker, DSO, who, when I was writing *Challenger: The Life of a Survey Ship* (London: Hollis & Carter, 1957), provided me with many of the photographs used in this chapter and lent me his diary of the winter party.
2 Founded by Moravian missionaries in 1771, Nain developed into a thriving community during the nineteenth century and continues to be the most important community on the northern Labrador coast. Though the HBC first appeared in this part of Labrador in the late 1860s, the trading station at the time of *Challenger's* survey was in fact a Moravian post which the HBC had leased from the missionaries in 1926. (Ed.)
3 The name remains today. (Ed.)

## CHAPTER SEVEN

1 *150 let Gidrograficheskoi sluzhbe Voenno-Morskogo flota*, 36–8.
2 *Zapiski po gidrografii*, 1938, 1:15.
3 Ibid., 1967, 4:34.
4 Ibid., 4:35.
5 RGAVMF, Fonds r-180, Opis [file list] 1, Delo [file] 65, 12.
6 Dikanskii, *Kalendar Kommunista na 1925 god*, 698.
7 *Zapiski po gidrografii*, 1938, 1:18.
8 Ibid., 1:19.
9 *Istoriia gidrograficheskoi sluzhby Rossiiskogo flota*, 1:358.
10 RGAVMF, Fonds 404, Opis 2, Delo 913, 136, 140, 142 verso.
11 Ibid., Delo 1098, 121–2.
12 Ibid., Delo 1276, 202 verso.
13 The first-class level was the most basic survey, and the third-class the most complete. (Ed.)
14 RGAMVF, Fonds r-552, Opis 1, Delo 24, 4–5.
15 RGAMVF, Fonds r-457, Opis 1, Delo 17, 1–4.
16 Ibid., 23.
17 RGAMVF, Fonds r-739, Opis 1, Delo 157, 32.
18 *Istoriia gidrograficheskoi sluzhby Rossiiskogo Flota*, 2:25.

19 Ibid., 1:360.
20 RGAVMF, Fonds 404, Opis 2, Delo 1278, 194–5.
21 Ibid., Delo 1486, 63–63 verso.
22 RGAMVF, Fonds r-457, Opis 1, Delo 17, 22 verso.
23 RGAMVF, Fonds r-180, Opis 1, Delo 80, 9.
24 *Zapiski po gidrografii*, 1938, 1:32.
25 Ibid., 1:35.
26 RGAVMF, Fonds r-180, Opis 1, Delo 405, 2 verso.
27 *Istoriia gidrograficheskoi sluzhby Rossiiskogo flota*, 2:28.
28 Ibid., 1:362–3.
29 RGAVMF, Fonds 404, Opis 2, Delo 1098, 161.
30 Ibid., Delo 1228, 10–11.
31 Ibid., 323.
32 Ibid., Delo 1789, 285.
33 *Istoriia gidrograficheskoi sluzhby Rossiiskogo flota*, 2:29.
34 RGAVMF, Fonds r-898, Opis 1, Delo 45, 19.
35 Ibid., 20–1.
36 *Zapiski po gidrografii*, 1938, 1:38.
37 Ibid., 1:39.
38 RGAVMF, Fonds r-180, Opis 1, Delo 80, 9–9 verso.
39 RGAMVF, Fonds r-739, Opis 1, Delo 138, 28.
40 *Zapiski po gidrografii*, 1938, 1:44.
41 Ibid., 1:45.

## CHAPTER EIGHT

1 Hayes, *Historical Atlas of the North Pacific Ocean*; Crane, *Mercator.*
2 Hayes, *Historical Atlas of the North Pacific Ocean*, 78.
3 Ibid., 93. My translation of the German heading, which was published in both German and French.
4 Hadley, *U-Boats against Canada*; Hadley and Sarty, *Tin-Pots and Pirate Ships.*
5 BA-MA, RM5/886, Immediatvortrag, 4 April 1904.
6 BA-MA, RM5/5343, vol. 1, "U-Plätze Amerika, Okt. 5–Nov. 6"; SMS *Panther*, 16 September 1905; also vol. 2, BA-MA: RM 5/5344, "U-Plätze Amerika, Jan 07 – Juli 13."
7 "Allerhöchste Belobigung" (Allerhöchste Kabinettsordre), 15 Oktober 1907; see BA-MA, RM5/5715, "England, Kolonien in Nordamerika."

8 Hadley and Sarty, *Tin-Pots and Pirate Ships*, 131–56; Messimer, *The Merchant U-Boat*.
9 Hadley and Sarty, *Tin-Pots and Pirate Ships*, 157–78.
10 Ibid., 233–62, 263–89.
11 BA-MA, RM5/6439, "U.K. Verband, Kabelschneide- und Minenangelegenheiten, Juli 1917 – August 1918"; Hadley and Sarty, *Tin-Pots and Pirate Ships*, 287.
12 Hadley and Sarty, *Tin-Pots and Pirate Ships*, 243.
13 BA-MA, Nr. 299, *Ubootshandbuch der Ostküste Kanadas, Sammelband* (1942), and *Ubootshandbuch der Ostküste Kanadas, Atlas* (1942).
14 BA-MA, RMD 5/2065, *Handbuch der Ostküste der Vereinigten Staaten von Nordamerika*, Teil I: *Von West Quoddy Head bis Sandy Hook*, Erste Auflage, Abgeschlossen mit "Nachrichten für Seefahrer" Ausgabe vom 3, Dezember 1942 (Berlin: E.S. Mittler und Sohn, 1943); RMD 5/2064, Teil II: *Von Sandy Hook bis Cape Canaveral*, Erste Auflage, Abgeschlossen mit "Nachrichten für Seefahrer" Ausgabe 28 vom 3, Juli 1941.
15 BA-MA, M. Dv. Nr 299, *Ubootshandbuch der Ostküste Kanadas*, Heft I: *Nordost- und Südostküste von Cape Breton Island, Südostküste von Nova Scotia, Fundy-Bucht*, Herausgegeben vom Oberkommando der Kriegsmarine (Berlin, August 1942); Heft II: *Neufundland und Belle Isle Strasse*, Herausgegeben vom Oberkommando der Kriegsmarine (Berlin, September 1942); Heft III: *St. Lorenz-Golf*, Herausgegeben vom Oberkommando der Kriegsmarine (Berlin, Oktober 1943).
16 Hadley, *U-boats against Canada*, 52–81.
17 Ibid., 144–67.
18 BA-MA, PG-30122, KTB/132 [war diary, *U-132*].
19 See my "Eight Modern Charts," in the Canadian War Museum, Ottawa, accession no. 20030129-001.
20 Milner, *The U-Boat Hunters*, 17–18.
21 See "SQ 36" in "Operations North American Waters – Gulf of Saint Lawrence," together with associated ship files, in the Directorate of History and Heritage, National Defence Headquarters, DHH: 1650–239/16.
22 Milner, *The U-boat Hunters*, 178.
23 Ibid.
24 For an exhaustive treatment of the German Naval Weather Service in the Arctic region, see Franz Selinger, *Von "Nanok" bis "Eismitte."*

25 BA-MA, PG 30569–576, KTB/537 [war diary, *U-537*]; see also Douglas, "The Nazi Weather Station in Labrador"; Douglas and Seliger, "Oktober 1943 – Juli 1981." William Glover forwarded to me information on German Hydrographic Institute charts 460FL and 461FL and their Canadian counterparts, LC 8047 and LC 8048.
26 Seliger, *Von "Nanok" bis "Eismitte,"* 308.
27 Kurt Petersen, to the author, quoted in my *U-Boats against Canada*, 227–8. See also Milner, *The U-Boat Hunters*, 200.
28 My translation. For the complete German text and my translation, see my *U-Boats against Canada*, 307–11. (The original is the better doggerel of the two.)
29 DHH, NSS 1048-48-31, A/Captain D.K. Laidlaw, RCN(T), director of Operations Division, to CNS, "Protection of shipping in the Canadian Coastal Zone," 21 November 1944.
30 Hadley, *U-Boats against Canada*, 292; McLean, "Muddling Through."
31 Hadley and Weber, "Einsatz und Abwehr von Schnorchel Ubooten vor Kanada 1944/45," 27.

## CHAPTER NINE

1 McGrath and Sebert, *Mapping a Northern Land*, 101–3.
2 Fillmore and Sandilands, *The Chartmakers*, 242–3.
3 A harpoon sounder was a coarse-pitched propeller mounted on the lead-line just above the weight. A brake would stop it turning when the lead hit the bottom. Geared wheels connected to the propeller would record the sounding.
4 Doekes, Bryant, and Macdonald, "Automated Selection of Bathymetric Data to Accurately Represent Bottom Topography."
5 Kerr, "A Production/Development Survey."
6 Bryant, "Side Scan Sonar for Hydrography."
7 Burke, "The Canadian Hydrographic Service (Atlantic) Sweep Program."
8 Burke, Forbes, and White, "Processing 'Large' Data Sets from 100% Bottom Coverage 'Shallow' Water Sweep Surveys."
9 Du, Wells, and Mayer, "Gridding High Volume High Density Bathymetric Data."
10 Kielland and Hally, "Canadian Preparations for the Swath Sounding Era."

11 Hare, "Depth and Position Error Budgets for Multibeam Echo-Sounding."
12 Crawford, "A Technique for Quality Control and Selection of Tidal Harmonic Constituents."
13 Tait and Ku, "A Historical Review of Tidal, Current and Water Level Surveying in the Canadian Hydrographic Service."
14 Dohler, *Construction of Tide Stations at Brevoort Harbour and Resolute Northwest Territories 1957*.
15 St. Jacques, "A Permanent Gauging System for Arctic Application."
16 Knudsen, "A New Water Level Acquisition and Telemetry System."
17 Stephenson and Silver, "The Low Power Tide Gauge."
18 Kielland and Hally, "Digital Tide Gauge – New Orientations."
19 Dohler, "The Mean Water Level at Pointe-au-Père as Used for International Great Lakes Datum (1955)."
20 Dohler, "The Tsunami Warning System in the Pacific."
21 Rutley, "The Demolition of Ripple Rock."
22 Personal communication to the author, Rapatz, 3 January 2003.
23 Freeman, Haras, and Wigen, "Hydrodynamic Surveys and the New Technology."
24 Crawford and Huggett, "The Tidal Streams Surrounding Vancouver Island."
25 Woodward and Huggett, "Performance of Deep Subsurface Current Meter Moorings."
26 Fillmore and Sandilands, *The Chartmakers*, 96–7.
27 MacPhee, Crutchlow, and Knudsen, "Arctic Hydrography – Past, Present and Future."
28 Douglas, "Hydrographic Research in the Canadian Arctic," 378–9.
29 Holladay et al., "A Quantitative Comparison of the Towed In-Flight Bathymetry System with Launch-Borne Acoustic Soundings near Coppermine NWT."
30 Douglas, "Remote Controlled Launch."
31 Malone, Burke, and Vine, "Dolphin: A Proven Hydrographic Vehicle."
32 Casey et al., "Airborne GPS Trials"; Hare, "Calibrating Larsen-500 Lidar Bathymetry in Dolphin and Union Strait."
33 The lack of understanding of losing a Decca lane was a contributing factor in the brand new *Baffin* running aground on Black

Rock, off Lunenburg, in 1957 (Fillmore and Sandilands, *The Chartmakers*, 193–4 and 212–13).

34 Johler, Kellar, and Walters, *Phase of the Low Radiofrequency Ground Wave.*

35 Brunavs and Wells, *Accurate Phase Lag Measurement over Seawater using Decca Lambda.*

36 Gray, "Propagation Velocity of Decca Frequency Transmissions over Sea Ice."

37 Miller, "Two Range Decca Installed on the Hydrographic Vessel *Kapuskasing* (Canada)"; LaCroix and Charles, "The Method and Use of Two-Range Decca."

38 Eaton, Mortimer, and Gray, "Accurate Chart Latticing for Loran-C."

39 Gray and Mortimer, "Loran-C Coordinate Converters – Plague or Panacea."

40 Eaton, "Experience with Hi-Fix Hyperbolic in the Canadian Arctic."

41 Canadian Hydrographic Service, *Annual Report*, 1973, 37.

42 Casey, "Improving the Performance of the Motorola MRS."

43 Janes, Eaton and Wilson, "UHF Syledis for Coastal Survey Positioning."

44 Fenn et al., "Self-Tracking Range-Bearing Positioning Systems."

45 Grant, "Rho-Rho Loran-C Combined with Satellite Navigation for Offshore Surveys."

46 Wells, "A New Integrated Navigation System: BIONAV."

47 For an example, see the difficulties in positioning Sable Island in Gray, "Where Has Sable Island Been for the Past 200 Years?"

48 Brookes, "Field Notes and Office Memos."

49 Shoran (Short Range Navigation) was a Second World War bombing aid that measured two distances. In its survey application, it used the two ranges in a line-crossing technique between ground points 130 to 600 kilometres apart.

50 Gray, "Ten Years of Experience in Converting Canadian Hydrographic Service Charts to a World-Based Geodetic System."

51 Macdonald, "Automation Today – Scratching the Twenty-One Year Itch."

52 Warren, Boone, and Guibord, "A Digital Depth Model and Automated Contours."

53 Meehan, "The Canadian Hydrographic Service," chapter 1.

54 Ibid., chapter 4a.

55 Kerr, "Surveying the sands off Sable," 22–4.
56 Meehan, "The Canadian Hydrographic Service," chapter 2.
57 Petticrew and Champ, "Development of a New Canadian Chart Presentation."
58 Covey, "Number 8."
59 Boyle, "Automation in Hydrographic Charting."
60 Evangelatos, "Automation in the Production of Nautical Charts."
61 Eaton, Anderson, and Evangelatos, "The Electronic Chart."
62 Eaton et al., "Learning from an Electronic Chart Testbed."
63 Vachon, "Electrostatic Printing: POD (Print-on-Demand) and Related Projects."
64 Holroyd, "Developing a Print on Demand Service in the Canadian Hydrographic Service."

## CHAPTER TEN

1 WHOI, DLA, William Metcalf Logs, USS *Edisto* – Arctic Cruise Spring 1951 – Greenland/Norwegian Sea. BT refers to the bathythermograph, an instrument deployed from the research vessel to make measurements of ocean temperature versus depth. The ammunition acquired on this voyage provided the embarked representatives of the Bureau of Ordnance with explosives for tests that formed part of the scheduled scientific activity.
2 WHOI, DLA, Institution Biography, William G. Metcalf.
3 Weir, *An Ocean in Common*, 210–18. In time, Fleming would take over the oceanography program at the University of Washington.
4 Ibid., 251–3.
5 WHOI, DLA, Metcalf Logs, USS *Edisto* – Arctic Cruise – Summer 1951; Long, *Ocean Sciences*, 93–5. The Military Sea Transportation Service (MSTS) was created by the US Department of Defense in 1950. See also WHOI, DLA, Metcalf Logs, Logs for Voyages with USS *Edisto* 1950–55, especially USS *Edisto* – Arctic Cruise – Baffin Bay – Thule, Greenland, 1 June to 10 July 1951.
6 WHOI, DLA, Metcalf Logs, USS *Edisto* – Arctic Cruise – Baffin Bay – Thule, Greenland, 1 June to 10 July 1951, introduction.
7 Long, *Ocean Sciences*, 95–6.
8 Weir, *An Ocean in Common*, 210–18.

9 Ibid., 258–64; interview with Dr Waldo Lyon by Gary E. Weir, 26 April 1994, in US Navy Operational Archive, US Naval Historical Center, Washington, DC; Leary, *Under Ice*. He conducted much of this early Arctic research, especially in terms of the operational applications of his work, in conjunction with his Canadian counterparts John P. Tully and William M. Cameron at the Canadian Biological Station at Nanaimo, BC.
10 WHOI, DLA, Metcalf Logs, Introduction, USS *Edisto* – Arctic Cruise – Spring 1951.
11 USS *Edisto*: Burton Island Class; 3,575 tons, capable of 16 knots, with a complement of 353.
12 US Naval Historical Center, Washington, DC, US Navy Operational Archive, Post 1 January Command File, Study of Undersea Warfare (Low Report), 22 April 1950; Weir, *Forged in War*, 133–9.
13 WHOI, DLA, Metcalf Logs, USS *Edisto* – Arctic Cruise Spring 1951. Subsequent quotations from Metcalf in the course of this cruise are from this source.
14 Weir, *An Ocean in Common*, chapter 11.
15 WHOI, DLA, Metcalf Logs, USS *Edisto* – Arctic Cruise Spring 1951.
16 Ibid.
17 Ibid.
18 Ibid.
19 Schlee, *The Edge of an Unfamiliar World*, 261–6.
20 WHOI, DLA, Metcalf Logs, Introduction, Skull Cliff, Alaska – 16 November 1952–26 February 1953 – Skull Cliff Loran Site. The "D" in the DEW acronym initially stood for "distant," but it later changed in some reports to "defense."
21 Ibid.
22 Ibid.
23 Ibid.

# *Bibliography*


## ARCHIVAL SOURCES

Archives nationales (Paris), Archives de la Marine
Archivo del Museo Naval, Madrid
British Columbia Archives
British Columbia Crown Lands Registry
Bundes- und Militärarchiv, Freiburg im Breisgau
Directorate of History and Heritage, National Defence Headquarters, Ottawa
National Archives of Canada
Oregon Historical Society, Malaspina Papers
Public Record Office, London
Rossiiskii gosudarstvennyi arkhiv Voenno-Morskogo Flota (Russian State Archives of the Military Fleet)
United Kingdom Hydrographic Office, Taunton
United States Navy Operational Archive, US Naval Historical Center, Washington, DC
Woods Hole Oceanographic Institution, Data Library and Archives

## ORAL HISTORY

Interview with Dr Waldo Lyon by Gary E. Weir, 26 April 1994
Study of Undersea Warfare (Low Report), 22 April 1950, Post 1 January Command File

PUBLISHED SOURCES

Akrigg, G.P.V., and H.B. Akrigg, *H.M.S. Virago in the Pacific 1851–1855: To the Queen Charlottes and Beyond.* Victoria: Sono Nis Press, 1992

*The American West Coast and Alaska: Original Drawings and Maps from the Expedition to Nootka Sound of Juan Francisco de la Bodega y Quadra, 1792.* Catalogue no. 144. New York: H.P. Kraus [n.d.]

Andreae, Christopher. *Lines of Country: An Atlas of Railway and Waterway History in Canada.* Toronto: Boston Mills Press, 1997

Angus, James. *A Respectable Ditch: A History of the Trent-Severn Waterway, 1833–1920.* Kingston: McGill-Queen's University Press, 1988

Appleton, Thomas E. *Usque ad Mare: A History of the Canadian Coast Guard and Marine Services.* Ottawa: Department of Transport, 1968

*Atlas maritime, ou recueil des cartes réduites dressées au dépôt des cartes et plans de la Marine pour le service des vaisseaux du roi.* Par ordre du ministère. Paris, 1751

*Atlas para el viage de las goletas Sutil y Mexicana al reconimiento del estrecho de Juan de Fuca en 1792.* Madrid, 1802

Barris, Theodore. *Fire Canoe: Prairie Steamboat Days Revisited.* Toronto: McClelland and Stewart, 1977

Bedini, Silvio A. *The Pulse of Time: Galileo Galilei, the Determination of Longitude and the Pendulum Clock.* Bibliotecca di Nuncius. Firenze: Leo S. Olschki, 1991

Belcher, Sir Edward. *Narrative of a Voyage round the World, Performed in Her Majesty's Ship Sulphur, during the Years 1836–1842.* 2 vols. London, 1843; facsimile ed., Folkestone: Dawsons, 1970

Bellin, J.N. "Remarques sur les cartes du 'Neptune françois,' dont les planches ont été remise au dépôt des plans de marine, en 1753." *Le Neptune françois.* Paris. (many editions)

Bennett, Gordon. *Yukon Transportation: A History.* Occasional Papers in Archaeology and History, no. 19. Ottawa: National Historic Parks and Sites Branch, Parks Canada, Indian and Northern Affairs, 1978

Berton, Pierre. *The National Dream.* Toronto: McClelland & Stewart 1970

Black, W.A. *Extracts Relating to the Navigability of Canadian Inland Waterways.* Canada Geographical Branch Paper no. 7. Ottawa: Department of Mines and Technical Surveys, 1957

Boyle, A.R. "Automation in Hydrographic Charting." *Canadian Surveyor* 24 (1970): 519–37
Brookes, W.R. "Field Notes and Office Memos." *Geomatica* 52 (1998): 246–7
Brunavs, P., and D.E. Wells. *Accurate Phase Lag Measurements over Seawater using Decca Lambda*. Atlantic Oceanographic Laboratory Report no. 1971–2. Halifax, 1971
Bryant, R.S. "Side Scan Sonar for Hydrography." *International Hydrographic Review* 52 (1975): 43–56
Burke, R. "The Canadian Hydrographic Service (Atlantic) Sweep Program – A Status Report." *Lighthouse* 27 (1983): 25–30
Burke, R.G., S. Forbes, and K. White. "Processing 'Large' Data Sets from 100% Bottom Coverage 'Shallow' Water Sweep Surveys." *International Hydrographic Review* 65 (1988): 75–89
Bush, Edward. *The Canadian Lighthouse*. Occasional Papers in Archaeology and History, no. 9. Ottawa: Parks Canada, Indian Affairs and Northern Development, 1974
Canada. Department of Fisheries and Oceans. *Sailing Directions: Great Lakes*. Vol. 2. Ottawa, 1984
– *Sailing Directions: Labrador and Hudson Bay*. 6th ed. Ottawa, 1988
Canada. Department of Marine and Fisheries. *Annual Report*[s]. Ottawa. Various years.
– *Canada: Her Natural Resources, Navigation, Principal Steamer Lines and Trans-Continental Railways*. Ottawa: Government Printing Bureau, 1912.
– *Georgian Bay and North Channel Pilot*. 1899.
Canada. Department of Public Works. *Annual Report*. Various years, 1868–1910
– *General Report, 1867; 1867–1882*
– "Georgian Bay Ship Canal Report." Canada, *Sessional Papers*. 1909, no. 19a
– *Reports and Documents in Reference to the Location of the Line and a Western Terminal Harbour*. Ottawa, 1878
– "Report on the Hudson Bay Railway Project." Canada, *Sessional Papers*, 1910, no. 20d
– "Report of a Commission Appointed to Inquire into the Cause of the Floods which Occur Periodically in the River St. Lawrence between Montreal and Quebec (1873)." In *General Report, 1867–1882*
– "Nelson River Report." Canada, *Sessional Papers*, 1911, no. 19b

– "Report to Minister of Public Works by the Commissioners on the Floods in the St. Lawrence River (1886)." Canada, *Sessional Papers*, 1889, no. 76

Canada, Department of Railways and Canals. *Annual Report*. Various years, 1879–1910

Canada. Department of the Interior. *Annual Report*. 1885

Canada. House of Commons. *Debates*. Ottawa. Various years

– "Reports of the Select Standing Committee … in Connection with the Tenders and Contracts respecting the Quebec Harbour Works and the Esquimalt Graving Dock and also Relative to the Resignation of Honorable Thomas McGreevy with Minutes of Proceedings and Evidence Attached." *Journals of the House of Commons*, appendix, vol. 25, 1891

Canada. Royal Commission. *Report of the Canadian Pacific Railway Royal Commission*. Vol 3: *Conclusions*. Ottawa, 1882

Canadian Hydrographic Service. *Annual Report, 1973*. Ottawa, 1974

Casey, M.J. "Improving the Performance of the Motorola MRS." *Lighthouse* 24(1981): 5–9

– "Multiranging – An Expensive Solution to a Simple Problem." *Lighthouse* 25 (1982): 41

– "Range Accuracy and Signal Strength in a Motorola Mini-Ranger System." *Lighthouse*, special ed. (1981): 37–49

– N. Beck, J.R. MacDougall, and R.A. O'Neil. "Airborne GPS Trials." *Lighthouse* 28 (1983): 19–21

– and D. Monahan. "Contours and Contouring in Hydrography – Part II." *Lighthouse* 30 (1984): 30–7

Cerezo, Martínez Ricardo, ed. *La Expedición Malaspina 1789–1794*. Tomo II, *Diario general del viaje por Alejandro Malaspina*. Madrid, 1990

Chabert de Cogolin, Joseph. *Voyage fait par ordre du roi en 1750 et 1751, dans l'Amérique septentrionale, pour rectifier les cartes des côtes de l'Acadie, de l'isle Royale & de l'isle de Terre–Neuve; et pour en fixer les principaux points par des observations astronomiques*. Paris, 1753; repr., New York, 1966

Chambers, Captain Ernest J. *The Canadian Marine: A History of the Department of Marine and Fisheries*. Toronto: Canadian Marine and Fisheries, 1905

Chappell, Edwin, ed. *The Tangier Papers of Samuel Pepys*. London: Navy Records Society, 1935

Conlon, P.M. "La Condamine the inquisitive." In *Studies on Voltaire and the Eighteenth Century* 55. Genève: Institut et Musée Voltaire, 1967

Cook, Andrew S. "Alexander Dalrymple (1737–1808), Hydrographer to the East India Company and to the Admiralty, as Publisher: A Catalogue of Books and Charts." PhD dissertation, University of St. Andrews, 1992

Cook, James, and James King. *A Voyage to the Pacific Ocean … for Making Discoveries in the Northern Hemisphere.* 3 vols. London, 1784

Covey, Bill. "Number 8." *Lighthouse* 42 (1990): 11–16

Cowie, Frederick W. *Report of the Honourable Frank Cochrane, Minister of Railways and Canals, on Halifax Harbour and the Development of a Project of Modern Ocean Terminals.* Ottawa: Department of Railways and Canals, 1913

Crane, Nicholas. *Mercator: The Man Who Mapped the Planet.* London: Weidenfeld & Nicolson, 2002

Crawford, William R. "A Technique for Quality Control and Selection of Tidal Harmonic Constituents." *International Hydrographic Review* 72 (1995): 135–50

– and W.S. Huggett. "The Tidal Streams Surrounding Vancouver Island." *Lighthouse* 23 (1981): 10–16

Cunnungham, Brysson. *A Treatise on the Principles and Practice of Harbour Engineering.* London: Charles Griffin, 1908

David, Andrew, Felipe Fernández-Armesto, Carlos Novi, and Glyn Williams, eds. *The Malaspina Expedition 1789–1794: Journal of the Voyage by Alejandro Malaspina.* Vol. 1. London: The Hakluyt Society, 2001

Dawson, L.S. *Memoirs of Hydrography including Brief Biographies of the Principal Officers Who Have Served in H.M. Naval Surveying Service between the Years 1750 and 1885.* 2 vols. Eastbourne: H.W. Keay, 1885; repub. in 1 vol. with preface and errata, London: Cornmarket Press, 1969

Day, Sir Archibald. *The Admiralty Hydrographic Service, 1795–1919.* London: HMSO, 1967

Delanglez, J., ed. "Journal de Louis Jolliet allant à la découverte du Labrador, 1694." *Rapport de l'archiviste de la province de Québec, 1943–1944.* Québec, 1944

Denis, Leo G. *Electric Generation and Distribution in Canada.* Ottawa, 1918 de Tremaudan, A.H. *The Hudson Bay Road*. Toronto: J.M. Dent, 1915

*Dictionary of Canadian Biography*. Vols 2–5, 13. Toronto: University of Toronto Press

Dikanskii, S.S., ed. *Kalendar Kommunista na 1925 god* [Calendar of the Communist for 1925]. Leningrad, 1925

Dixon, George. *A Voyage round the World ... in the King George and Queen Charlotte*. London: Ge. Goulding, 1789

Doekes, C., R.S. Bryant, and G. Macdonald. "Automated Selection of Bathymetric Data to Accurately Represent Bottom Topography." *Proceedings of the 15th Annual Canadian Hydrographic Conference*. Ottawa, 1976

Dohler, G.C. *Construction of Tide Stations at Brevoort Harbour and Resolute Northwest Territories 1957*. Ottawa: Surveying and Mapping Branch, Department of Mines and Technical Surveys, 1958

– "The Mean Water Level at Pointe-au-Père as Used for International Great Lakes Datum (1955)." *International Hydrographic Review* 48 (1971): 117–28

– "The Tsunami Warning System in the Pacific." *Lighthouse* 30 (1984): 3–7

Douglas, Ross. "Hydrographic Research in the Canadian Arctic." *Canadian Surveyor* 33 (1979): 378–80

– "Remote Controlled Launch." *Canadian Surveyor* 23 (1969): 62

Douglas, W.A.B. "The Nazi Weather Station in Labrador." *Canadian Geographic* 101 (Dec. 1981/Jan. 1982): 42–7

– and Franz Seliger, "Oktober 1943 – Juli 1981: Eine Marine-Wetterstation auf Labrador." *Marine-Rundschau*, Nr. 5 (Mai 1982): 256–62

Du, Ziqin, David Wells, and Larry Mayer. "Gridding High Volume High Density Bathymetric Data." *Geomatica* 50 (1996): 149–58

Eaton, R.M. "Experience with Hi-Fix Hyperbolic in the Canadian Arctic." *Supplement to the International Hydrographic Review* 6 (1965): 39–49

– N.M. Anderson, and T.V. Evangelatos. "The Electronic Chart." *Proceedings, Centennial Conference Canadian Hydrographic Service* (1983), 188–95

– H. Astle, S.J. Glavin, S.T. Grant, S.E. Masry, and B.W. Shaw. "Learning from an Electronic Chart Testbed." *International Hydrographic Review* 67 (1990): 31–43

– A.R. Mortimer, and D.H. Gray. "Accurate Chart Latticing for Loran-C." *International Hydrographic Review* 56 (1979): 21–33

Espinosa y Tello, José de. *Memoria sobre las observaciones astronónicas, que han servido de fundamento á las cartas de la costa NO de América, publicadas por la direccion de trabajos hidrográficos, á continuacion del viage de las goletas Sutil y Mexicana al estrecho de Juan de Fuca*. Madrid, 1805

– *Memorias sobre las observaciones astronómicos hechas por los navegantes españoles en distintos lugares del globo*. 2 vols. Madrid, 1809

Evangelatos, Timothy V. "Automation in the Production of Nautical Charts." *Canadian Surveyor* 31 (1977): 4–8

Fenn, G., P. Millette, B. Tinney, and D. St. Jacques. "Self-Tracking Range-Bearing Positioning Systems – An Evaluation." *Lighthouse* 38 (1988): 21–34

Fillmore, Stanley, and R.W. Sandilands. *The Chartmakers: A History of Nautical Surveying in Canada*. Toronto: NC Press, 1983

Foster, Michael, and Carol Marino. *The Polar Shelf: The Saga of Canada's Arctic Scientists*. Toronto: NC Press, 1986

Freeman, Nelson, W.S. Haras, and S.O. Wigen. "Hydrodynamic Surveys and the New Technology." *Canadian Surveyor* 28:3 (1974): 225–36

Gauvin, Michel. "The Reformer and the Machine: Montreal Civic Politics from Raymond Préfontaine to Médéric Martin." *Journal of Canadian Studies* 13, no. 2 (summer 1978): 16–26

Gibson, James R. *Otter Skins, Boston Ships, and China Goods: The Maritime Fur Trade of the Northwest Coast, 1785–1841*. Montreal: McGill-Queen's University Press, 1992

Gimblett, Richard H. "Reassessing the Dreadnought Crisis of 1909 and the Origins of the Royal Canadian Navy." *The Northern Mariner /Le marin du nord* 4, no. 1 (January 1994): 35–53

Glover, William. "The Challenge of Navigation to Hydrography on the British Columbia Coast, 1850–1930." *The Northern Mariner/ Le marin du nord* 6, no. 4 (1996): 1–16

– "Early Trading Voyages in Hudson Bay, 1700–1750." In Nicholas Tracy, ed., *The Age of Sail: The International Annual of the Historic Sailing Ship*. London: Conway Maritime Press, 2002

Gonzáles, Francisco José. "Instrumentos del Real Observatorio destinados a la Expedició Malaspina." In *Malaspina '92 Jornados Internacionales Cádiz – 1994*. Cádiz, 1994

Gough, Barry M. *Gunboat Frontier: British Maritime Authority and Northwest Coast Indians, 1846–1890*, Vancouver: University of British Columbia Press, 1984

– *The Royal Navy and the Northwest Coast of North America, 1810–1914: A Study of British Maritime Ascendancy.* Vancouver: University of British Columbia Press, 1971

Grant, S.T. "Rho-Rho Loran-C Combined with Satellite Navigation for Offshore Surveys." *International Hydrographic Review* 50 (1973): 35–54

Gray, David H. "Propagation Velocity of Decca Frequency Transmissions over Sea Ice." *Canadian Surveyor* 29 (1975): 277–88

– "Ten Years of Experience in Converting Canadian Hydrographic Service Charts to a World-Based Geodetic System." *International Hydrographic Review* 75 (1988): 13–26

– "Where Has Sable Island Been for the Past 200 Years?" *CISM Journal* 46 (1992): 265–75

– and Tony Mortimer. "Loran-C Coordinate Converters – Plague or Panacea." *International Hydrographic Review* 59 (1982): 109–18

Gray, N.G. "Reminiscences of a Retired Hydrographer Lake Superior." *Lighthouse* 28 (1983): 41–4

Great Britain. Colonial Office. *Conference between the Secretary of State for the Colonies and the Premiers of Self-Governing Colonies (Miscellaneous no. 144): Minutes of Proceedings and Papers Laid Before the Conference.* NA, R6537-0-1-E [formerly MG 27, 1J17], reel C-1366/67

Hadley, Michael L. *U-Boats against Canada: German Submarines in Canadian Waters.* Montreal and Kingston: McGill-Queen's University Press, 1985, 1990

– and Roger Sarty. *Tin-Pots and Pirate Ships: Canadian Naval Forces and German Sea Raiders.* Montreal and Kingston: McGill-Queen's University Press, 1991

– and Fritz Rudolf Weber. "Einsatz und Abwehr von Schnorchel Ubooten vor Kanada 1944/45." In Ludger Steenpaß, ed., *Seekriegsoperationen im Zweiten Weltkrieg – ausgewählte Beispiele und Beiträge von Zeitzeugen.* 13–62. Historisch-Taktische Tagung der Flotte. Glücksburg: Marineunterstützungskommando, 1987

Hahn, Roger. *The Anatomy of a Scientific Institution: The Paris Academy of Sciences, 1666–1803.* Berkeley and Los Angeles: University of California Press, 1971

– "Les observatoires en France au XVIIIe siècle." In René Taton, ed., *Enseignement et diffusion des sciences en France au XVIIIe siècle.* Paris: Hermann, 1964

Hare, Rob. "Calibrating Larsen-500 Lidar Bathymetry in Dolphin and Union Strait Using Dense Acoustic Ground-Truth." *International Hydrographic Review* 71 (1994): 91–108

– "Depth and Position Error Budgets for Multibeam Echo-Sounding." *International Hydrographic Review* 72:2 (1995): 37–70

Haycock, Ronald. *Sam Hughes: The Public Career of a Controversial Canadian, 1885–1916.* Canadian War Museum Historical Publication, no. 21. Waterloo: Wilfrid Laurier University Press in collaboration with Canadian War Museum, 1986

Hayes, Derek. *Historical Atlas of British Columbia and the Pacific Northwest.* Vancouver: Cavendish Books, 1999

– *Historical Atlas of Canada.* Vancouver: Douglas & McIntyre, 2002

– *Historical Atlas of the North Pacific Ocean: Maps of Discovery and Scientific Exploration, 1500–2000.* Vancouver and Toronto: Douglas & McIntyre, 2001

Heidenreich, C.E. *Explorations and Mapping of Samuel de Champlain, 1603–1632.* Cartographica, Monograph no. 17. Toronto: B.V. Gutsell, 1976

Hind, Henry Youle. *Narrative of the Canadian Red River Exploring Expedition of 1857 and of the Assiniboine and Saskatchewan Exploring Expedition of 1858.* London, 1860; repr. New York: Greenwood Press, 1969

*Historical Atlas of Canada.* Geoffrey J. Matthews, cartographer/designer. 3 vols. Toronto: University of Toronto Press, 1987–93

Holladay, J. Scott, David Wright, Michael R. Crutchlow, and Albert Koudys. "A Quantitative Comparison of the Towed In-Flight Bathymetry System with Launch-Borne Acoustic Soundings near Coppermine NWT." *Lighthouse* 49 (1994): 19–23

Holroyd, Paul. "Developing a Print on Demand Service in the Canadian Hydrographic Service." *International Hydrographic Review* 1, no. 2 (2000): 24–9

Hopkins, J. Castell. *The Canadian Annual Review of Public Affairs.* 1902–05. Toronto: The Annual Review Publishing Co., 1903–06

Howay, F.W. "A List of Trading Vessels in the Maritime Fur Trade, 1785–1794 [1795–1804; 1805–1814; 1815–1819; 1820–1825]." *Transactions of the Royal Society of Canada,* 3rd ser., 24 (1930), section 2, 111–34; 25 (1931), section 2, 117–49; 26 (1932), section 2, 43–86; 27 (1933), section 2, 119–47; 28 (1934), section 2, 11–49. Republished as *A List of Trading Vessels in the Maritime Fur*

*Trade, 1785–1825*, ed. R.A. Pierce (Materials for the Study of Alaskan History, 2; Kingston: Limestone Press, 1973)

Howse, Derek, and Beresford Hutchinson. *The Clocks and Watches of Captain James Cook, 1769–1969*. London: Antiquarian Horological Society [1970?]

Hydrographic Office, Admiralty. *The British Columbia Pilot. Including the Coast of British Columbia, from Juan de Fuca Strait to Portland Canal, together with Vancouver and Queen Charlotte Islands. Compiled from Admiralty Surveys*. [1st ed.]. London: HMSO, 1888. With *Hydrographic Notice* [1891] and *Supplement* 1895. 2d ed. (London: HMSO, 1898), with *Supplement* 1899 and *Revised Supplement* 1903. 3rd ed., (London: HMSO, 1905), with *Supplement* 1908 and *Revised Supplement* 1910. Subsequently (1913) issued in two parts, each part thereafter issued separately in an ongoing series of further editions

– *Catalogue of Charts, Plans, Views, and Sailing Directions, &c. published by order of the Lords Commissioners of the Admiralty*. London: Admiralty Office [later HMSO], 1825– Issues (with variations in title) from 1825 onwards.

– *Instructions to be followed in Writing and Editing Sailing Directions and in Remark Books* ... London: HMSO, 1857–

– *Ocean Passages for the World*. 2nd ed. Revised by Commander A.F.B. Woodhouse, RN. 1950.

– *Queen Charlotte Islands on Western Coast of North America: Remarks for Sailing Directions, Made in the Year 1853, by George H. Inskip, Master, R.N., H.M. Steam Vessel Virago*. London: HMSO, 1856

– *Vancouver Island Pilot, Part I: Sailing Directions for the Coasts of Vancouver Island and British Columbia, from the Entrance of Juan de Fuca Strait to Burrard Inlet and Nanaimo Harbour, by Captain George Henry Richards, R.N*. London: HMSO, 1861

– *The Vancouver Island Pilot: Containing Sailing Directions for the Coasts of Vancouver Island, and part of British Columbia. Compiled from the Surveys made by Captain George Henry Richards, R.N., in H.M. Ships Plumper and Hecate, between the years 1858 and 1864*. London: HMSO, 1864. With *Hydrographic Notices, Vancouver Island Pilot, Notice No. 1* to *No. 7* (London: HMSO, 1867–83)

– *Vancouver Island Pilot: Supplement: Coast of British Columbia from Queen Charlotte Sound to Portland Canal, including Queen Charlotte Islands. Compiled chiefly from Admiralty Surveys* London: HMSO, 1883

*Istoriia gidrograficheskoi sluzhby Rossiiskogo flota* [History of the Hydrographic Service of the Russian Fleet]. St Petersburg, 1997

Janes, H.W., R.M. Eaton, and J. Wilson. "UHF Syledis for Coastal Survey Positioning." *International Hydrographic Review* 62, no. 2 (1985): 59–73

Johler, J.R., W.J. Kellar, and L.C. Walters. *Phase of the Low Radio-frequency Ground Wave*. National Bureau of Standards Circular. 1956

Jones, T.B. *The Figure of the Earth*. Lawrence, Kan.: Coronado Press, 1967

Jouan, René. *Histoire de la Marine française*. Paris, 1950

Keefer, T.C. "Ice Floods and Winter Navigation of the Lower St. Lawrence." *Transactions of the Royal Society of Canada*, 2d ser., 4 (1898–99), section 3, 3–17

Kendrick, John. *Alejandro Malaspina: Portrait of a Visionary*. Montreal: McGill-Queen's University Press, 1999

– trans. and ed. *The Voyage of Sutil and Mexicana, 1792: The Last Spanish Exploration of the Northwest Coast of America*. Spokane: Arthur H. Clark Co., 1991

Kerr, A.J. "A Production/Development Survey – The Lower St. Lawrence River Survey, 1969." *Proceedings of the Ninth Annual Canadian Hydrographic Conference*, Ottawa, 1970

– "Surveying the Sands off Sable." *Lighthouse* 27 (1983): 22–4

Kielland, P., and P. Hally. "Canadian Preparations for the Swath Sounding Era." *Lighthouse* 39 (1989): 25–33

– "Digital Tide Gauge – New Orientations." *Proceedings of the 1989 Canadian Hydrographic Conference*, 214–22

Knudsen, D.A. "New Water Level Acquisition and Telemetry System." *Proceedings of the 15th Annual Canadian Hydrographic Conference*. Ottawa, 1976

LaCroix, G.W., and D'A.H. Charles. "The Method and Use of Two-Range Decca." *Supplement to the International Hydrographic Review* 1 (1960): 9–31

Lamb, W. Kaye, ed. *George Vancouver: A Voyage of Discovery to the North Pacific Ocean and round the World 1791–1795*. 4 vols. London: Hakluyt Society, 1984

LaPierre, L.L. "Joseph Israël Tarte and the McGreevy-Langevin Scandal." *Canadian Historical Association Report*, 1961, 47–57

Lavallée, Omer S.A. *Narrow Gauge Railways of Canada*. Montreal: Railfare, 1972

Leary, William M. *Under Ice: Waldo Lyon and the Development of the Arctic Submarine*. College Station: Texas A&M University Press, 1999

Legget, Robert. *Canals of Canada*. Vancouver: Douglas, David & Charles, 1976

Letourneau, Roger. "Grand Rapids Tramway: A Centennial History." *Beaver*, autumn 1977

Linteau, Paul-André. *Histoire de Montréal depuis la Confédération*. Montréal: Boréal, 1992

Long, John, ed. *Ocean Sciences*. Annapolis, Md: Naval Institute Press, 1964

Longstaff, F.V. *Esquimalt Naval Base: A History of Its Work and Its Defences*. Victoria: Victoria Book & Stationery Co., 1941

– and W. Kaye Lamb. "The Royal Navy on the Northwest Coast, 1813–1850." *British Columbia Historical Quarterly* 9 (1945): 1–24, 113–28

Lower, Arthur R.M. *Great Britain's Woodyard: British America and the Timber Trade, 1763–1867*. Kingston: McGill-Queen's University Press, 1973

Macdonald, George. "Automation Today – Scratching the Twenty-One Year Itch." *Proceedings, Centennial Conference Canadian Hydrographic Service*, 1983, 157–8

McGahan, Elizabeth. *The Port of Saint John: From Confederation to Nationalization, 1867–1927*. Saint John, NB: National Harbours Board, 1982

McGrath, Gerald, and Louis Sebert, eds. *Mapping a Northern Land: The Survey of Canada, 1947–1994*. Montreal and Kingston: McGill-Queen's University Press, 1999

McLean, Doug M. "Muddling Through: Canadian Anti-Submarine Doctrine and Practice, 1942–1945." In Michael L. Hadley, et al., eds., *A Nation's Navy: In Quest of Canadian Naval Identity*. Montreal and Kingston: McGill-Queen's University Press, 1996

MacPhee, S.B., M. Crutchlow, and D. Knudsen. "Arctic Hydrography – Past, Present and Future." *Lighthouse* 18 (1978): 3–5

Maginley, Charles D., and Bernard Collin. *The Ships of Canada's Marine Services*. St Catharines, Ont.: Vanwell Publishing, 2001

Malone, A.K., R.G. Burke, and R. Vine. "Dolphin: A Proven Hydrographic Vehicle." *Lighthouse* 30 (1984): 38–41

Meehan, O.M. "An Outline of Hydrography in Canada." Typescript, Canadian Hydrographic Service, 1964–66. Published in four parts in *Soundings* (the house journal of the Canadian Hydrographic Service), 3(1) for December 1964, 4(1) for April 1965, 5(1) for August 1966, and 5(2) for December 1966. Various derived typescript versions exist; part II is entitled "A Few Aspects of British Admiralty Hydrography in Canada, 1795–1949"

Melville, T.R. "Canada and Sea Power: Canadian Naval Thought and Policy, 1860–1910." PhD thesis, Duke University, 1981

Messimer, Dwight R. *The Merchant U-Boat: Adventures of the* Deutschland, *1916–1918*. Annapolis, Md: Naval Institute Press, 1988

Miller, Al. "Surveying with Radio and Acoustics." *Lighthouse* 43 (1991): 15

Miller, Ian A. "Two Range Decca Installed on the Hydrographic Vessel *Kapuskasing* (Canada)." *International Hydrographic Review* 34, no. 2 (1957): 19–28

Milner, Marc. *The U-Boat Hunters: The Royal Canadian Navy and the Offensive against Germany's Submarines*. Toronto: University of Toronto Press, 1994

Monahan, D., and M.J. Casey. "Contours & Contouring in Hydrography – Part I." *Lighthouse* 28 (1983): 10–17

Morton, Desmond. *Ministers and Generals: Politics and the Canadian Militia, 1868–1904*. Toronto: University of Toronto Press, 1970

Neuville, Didier. *État sommaire des archives de la Marine anterieur à la révolution*. 1898; Lichtenstein: Krause Reprint, 1977

Norrie, Kenneth, and Douglas Owram. *A History of the Canadian Economy*. 2nd ed. Toronto: Harcourt Brace, 1996

Ollivier, Maurice, comp. and ed. *The Colonial and Imperial Conferences from 1887 to 1937*. Vol I, *Colonial Conferences* [1887, 1894, 1897, 1902, and 1907]. Ottawa: Queen's Printer, 1954

Olmstead, J.W. "The Voyage of John Richer to Acadia in 1670: A Study in the Relations of Science and Navigation under Colbert." *Proceedings of the American Philosophical Society* 104, sect. 6 (1960): 612–34

*150 let Gidrograficheskoi sluzhbe Voenno-Morskogo flota. Istoricheskii ocherk* [150 years of the Hydrographic Service of the Naval Fleet. A historical sketch]. Leningrad, 1977.

Otter, A.A. den. "Irrigation and Flood Control." In *Building Canada: A History of Public Works*. Toronto: University of Toronto, 1988

Owram, Douglas. *Building for Canadians: A History of the Department of Public Works, 1840–1960*. Ottawa: Public Works, 1979

Palau, Mercedes, Aránzazu Zabala, and Blanca Sáiz, eds. *Viaje científico y político ... Diario de viaje de Malaspina*. Madrid, 1984

Passfield, Robert. "Waterways." In *Building Canada: A History of Public Works*. Toronto: University of Toronto, 1988

Peel, Bruce. *Steamboats on the Saskatchewan*. Saskatoon: Western Producer, 1972

Penlington, Norman. *Canada and Imperialism, 1896–1899*. Toronto: University of Toronto Press, 1965

Petticrew, R.F., and C.G. Champ. "Development of a New Canadian Chart Presentation." *Proceedings of the 15th Annual Canadian Hydrographic Conference*, 1976

Phipps, Constantine John. *A Voyage towards the North Pole: Undertaken by His Majesty's Command, 1773*. London: Printed by W. Bowyer and J. Nichols, for J. Nourse, 1774

Prelini, Charles. *Dredges and Dredging*. New York: Van Nostrand, 1911

Pritchard, J.S. "Early French Hydrographic Surveys in the Saint Lawrence River." *International Hydrographic Review* 56, no. 1 (January 1979): 125–42.

– "French Charting of the East Coast of Canada." In Derek Crouse ed., *Five Hundred Years of Nautical Science, 1400–1900*. Greenwich: National Maritime Museum, 1981

Quinn, Alonzo DeF. *Design and Construction of Ports and Marine Structures*. 2nd ed. New York: McGraw-Hill, 1972

*Relación del viage hecho por las goletas Sutil y Mexicana en al año 1792 para reconocer el estrecho de Juan de Fuca*. Madrid, 1802

*Relation de ce qui s'est passé de plus remarquable aux missions des pères de la Compagnie de Jesus en la Nouvelle France, les années 1670 et 1671*. Paris: Sebastien Mabre-Cramoisy, 1672

Richards, G.H. *A Memoir of the Hydrographical Department of the Admiralty, 1868*. London: HMSO, 1868

Ritchie, G.S. *The Admiralty Chart: British Naval Hydrography in the Nineteenth Century*. London: Hollis & Carter, 1967; new ed. Edinburgh: Pentland, 1995

Roos, Arnold. "Electricity." In *Building Canada: A History of Public Works*. Toronto: University of Toronto, 1988

Ruggles, R.I. *A Country So Interesting: The Hudson's Bay Company and Two Centuries of Mapping, 1670–1870*. Rupert's Land Record Society series, 2. Montreal: McGill-Queen's University Press, 1991

Russo, François. "L'enseignement des sciences de la navigation dans les écoles d'hydrographie au XVIIIe siècle." In M. Mollat, ed., *Le navire et l'économie maritime du nord de l'Europe du moyen-age au 18e siècle*. Paris, 1958

Rutley, J.A.A. "The Demolition of Ripple Rock." *International Hydrographic Review* 36 (1959): 1, 19–28

St. Jacques, D.A. "A Permanent Gauging System for Arctic Application." *Lighthouse* 34 (1984): 7–9

Sáiz, Blanca. *Bibliografia sobre Alejandro Malaspina: y acerca de la expedición Malaspina y de los marinos y científicos que en ella participaron*. Madrid: Edicions El Museo universal, 1992

Sandilands, R.W. "The History of Hydrographic Surveying in British Columbia." In B. Farrell and A. Desbarats, eds., *Explorations in the History of Canadian Mapping: A Collection of Essays*, 113–31. Ottawa: Association of Canadian Map Libraries and Archives, 1988. Reprinted from duplicate typescript, Canadian Hydrographic Service, 1965

– "Hydrographic Charting and Oceanography on the West Coast of Canada from the Eighteenth Century to the Present Day." *Proceedings of the Royal Society of Edinburgh* 74 (1971): 65–82

– "Hydrographic Surveying in the Great Lakes during the Nineteenth Century." *The Canadian Surveyor* 36:2 (June 1982): 139–63

Schlee, Susan. *The Edge of an Unfamiliar World*. New York: E.P. Dutton, 1973

Schull, Joseph. *Laurier: The First Canadian*. Toronto: Macmillan, 1965

Seliger, Franz. *Von "Nanok" bis "Eismitte": Meteorologische Unternehmungen in der Arktis 1940–1945*. Schriften des Deutschen Schiffahrtsmuseums, Band 53. Hamburg: Convent Verlag, 2001

Soble, Dava. *Longitude: The True Story of a Lone Genius Who Solved the Greatest Scientific Problem of His Time*. New York: Penguin Books [1995], 1996

Spittle, John D. *Maps Printed by the Royal Engineers and the Lands & Works Department, New Westminister, British Columbia, 1861–1866*. Vancouver, 19XX

Stephens, George W., and Frederick W. Cowie. "Report on British and Continental Ports, with a View to the Development of the Port of Montreal and Canadian Transportation." Canada, *Sessional Papers*, 1908, no. 21c

Stephenson, F. "An Assessment of the Permanent Water Level Stations in the Canadian Arctic." *Lighthouse* 16 (1977): 21–5

Stephenson, F.E., and D.D. Silver. "The Low Power Tide Gauge – A New Tool for Hydrography." *Proceedings of the 1989 Canadian Hydrographic Conference*, 223–9

Stuifbergen, N. "SYLEDIS – a New Medium-Range Positioning System." *Lighthouse* 18 (1978): 6

Tait, B.J., and L.F. Ku. "A Historical Review of Tidal, Current and Water Level Surveying in the Canadian Hydrographic Service." *Proceedings, Centennial Conference Canadian Hydrographic Service*, 1983, 51–61

Talbot, Frederick. *The Making of a Great Canadian Railway*. London: Seeley, Service & Co., 1912

Taylor, A.D. "The Seas in the Seventies." In A.M.J. Hyatt, ed., *Dreadnought to Polaris: Maritime Strategy since Mahan*. Papers from the Conference on Strategic Studies at the University of Western Ontario, March 1972. Toronto: Copp Clark Pub. co., 1973

Taylor, David. "Degree of Difficulty: Measuring the Earth Proved to Be a Triumphant Fiasco." *Mercator's World* 4, no. 3 (May/June 1999)

Thomson, Don W. *Men and Meridians: The History of Surveying and Mapping in Canada*. Vol 2. Ottawa: Energy, Mines & Resources, 1967

Tizard, T.H. *Chronological List of the Officers Conducting British Maritime Discoveries and Surveys; together with the Names of the Vessels Employed, from the Earliest Times until 1900*. Hydrographic Department, no. 174. London: HMSO, 1900

Toronto Harbour Commissioners. *Toronto Harbour: The Passing Years*. Toronto, 1985

Tramond, Johannes. *Manuel d'histoire maritime de la France des origines à 1815*. Nouv. éd. Paris: Société d'éditions géographiques, maritimes et coloniales, 1946

Tucker, Gilbert Norman. *The Naval Service of Canada: Its Official History*. Vol. I, *Origins and Early Years*. Ottawa: King's Printer, 1952

Turner, Robert D. *Sternwheelers and Steam Tugs: An Illustrated History of the Canadian Pacific Railway's British Columbia Lake and River Service*. Victoria: Sono Nis Press, 1984

Vachon, D. "Electrostatic Printing: POD (Print-On-Demand) and Related Projects." *Lighthouse* 36 (1986): 3–8

Vancouver, George. *A Voyage of Discovery to the North Pacific Ocean and round the World*. 3 vols. and an atlas. London, 1798

Vervoort, P. "Lakehead Terminal Elevators: Aspects of Their Engineering History." *Canadian Journal of Civil Engineering* 17 (1990): 404–12

Wagner, Henry R. *Cartography of the Northwest coast of America to the Year 1800*. 2 vols. Berkeley, 1937; repub. in 1 vol., Amsterdam: N. Israel, 1968

Walbran, J.T. *British Columbia Coast Names, 1592–1906: Their Origin and History.* Ottawa: Government Printing Bureau, 1909; facsimile ed., with introduction by G.P.V. Akrigg, Vancouver: J.J. Douglas for Vancouver Public Library, 1971

Wallace, W. Stewart, ed. *The Macmillan Dictionary of Canadian Biography.* 3rd ed. Toronto: Macmillan, 1963

Warren, J.S., L.G. Boone, and P. Guibord. "A Digital Depth Model and Automated Contours." *Lighthouse* 41 (1990): 15–18

Weir, Gary E. *Forged in War: The Naval Industrial Complex and American Submarine Construction, 1940–1961*. Washington, DC: Brasseys, 1998

– *An Ocean in Common: American Naval Officers, Scientists, and the Ocean Environment*. College Station: Texas A&M University Press 2001

Wells, D.E. "A New Integrated Navigation System: BIONAV." *Proceedings of the Symposium of Position Fixing at Sea*, 1980, 1–26

Wharton, Sir William J.L. *Hydrographic Surveying: A Description of Means and Methods Employed in Constructing Marine Charts.* 4th ed. London: John Murray, 1920

Woodward, M.J., and W.S. Huggett. "Performance of Deep Subsurface Current Meter Moorings." *Lighthouse*, special ed. (1980): 124–30

Woolf, H. *The Transits of Venus: A Study of Eighteenth-Century Science*. Princeton: Princeton University Press, 1959

*Zapiski izdavayemiya Gosudarstvennim Admiralteiskim Departmentom, otnosyashchiyasya k' Moryeplavaniyu, Naukam i Slovesnosti* (Proceedings of the State Admiralty Department relating to Navigation, Science, and Literature). Various years

*Zapiski po gidrografii* [Notes on hydrography]. Various years

# Index